Ceramics in Nuclear and Alternative Energy Applications

Ceramics in Nuclear and Alternative Energy Applications

A Collection of Papers Presented at the 30th International Conference on Advanced Ceramics and Composites January 22–27, 2006, Cocoa Beach, Florida

Editor
Sharon Marra

General Editors
Andrew Wereszczak
Edgar Lara-Curzio

A JOHN WILEY & SONS, INC., PUBLICATION

Published by John Wiley & Sons, Inc., Hoboken, New Jersey
Published simultaneously in Canada.

Library of Congress Cataloging-in-Publication Data is available.

ISBN-13 978-0-470-08055-9
ISBN-10 0-470-08055-8

10 9 8 7 6 5 4 3 2 1

Contents

Preface

In January 2006, The American Ceramic Society's (ACerS) Nuclear and Environmental Technology Division supported ACerS Engineering Ceramics Division with their sponsorship of The 30th International Conference and Exposition on Advanced Ceramics and Composites. The meeting was held January 22-27, 2006 in Cocoa Beach, Florida. A symposium on Ceramics in Nuclear and Alternative Energy Applications was held. This volume documents a number of the papers presented. The symposium was extremely well received and offered some diverse topics to this well established conference. Papers from several countries were presented on timely topics such as dealing with alternative energy sources and ceramics in nuclear applications.

The success of the symposium and the proceedings wouldn't have been possible without the support and guidance of the staff at The American Ceramic Society. Our thanks to the ACerS Engineering Ceramics Division for allowing us to partner with them in this successful meeting. We look forward to future partnerships. Also, thanks to the session chairs - Jim Marra, David Peeler, Connie Herman, Tim Burchell, and Lance Snead.

SHARON MARRA
YUTAI KATOH
DANE SPEARING

Preface

Introduction

This book is one of seven issues that comprise Volume 27 of the Ceramic Engineering & Science Proceedings (CESP). This volume contains manuscripts that were presented at the 30th International Conference on Advanced Ceramic and Composites (ICACC) held in Cocoa Beach, Florida January 22–27, 2006. This meeting, which has become the premier international forum for the dissemination of information pertaining to the processing, properties and behavior of structural and multifunctional ceramics and composites, emerging ceramic technologies and applications of engineering ceramics, was organized by the Engineering Ceramics Division (ECD) of The American Ceramic Society (ACerS) in collaboration with ACerS Nuclear and Environmental Technology Division (NETD).

The 30th ICACC attracted more than 900 scientists and engineers from 27 countries and was organized into the following seven symposia:

- Mechanical Properties and Performance of Engineering Ceramics and Composites
- Advanced Ceramic Coatings for Structural, Environmental and Functional Applications
- 3rd International Symposium for Solid Oxide Fuel Cells
- Ceramics in Nuclear and Alternative Energy Applications
- Bioceramics and Biocomposites
- Topics in Ceramic Armor
- Synthesis and Processing of Nanostructured Materials

The organization of the Cocoa Beach meeting and the publication of these proceedings were possible thanks to the tireless dedication of many ECD and NETD volunteers and the professional staff of The American Ceramic Society.

ANDREW A. WERESZCZAK
EDGAR LARA-CURZIO
General Editors

Oak Ridge, TN (July 2006)

Irradiation Effects in Ceramics

(GENIV) NEXT GENERATION NUCLEAR POWER AND REQUIREMENTS FOR STANDARDS, CODES AND DATA BASES FOR CERAMIC MATRIX COMPOSITES

Michael G. Jenkins
University of Detroit Mercy
Detroit, MI 48221 USA

Edgar Lara-Curzio
Oak Ridge National Laboratory
Oak Ridge, TN 37831 USA

William E. Windes
Idaho National Laboratory
Idaho Falls, ID 83415 USA

ABSTRACT

Innovative and novel applications of CMCs are critical to the success of GenIV: Next Generation Nuclear Power (NGNP) plants. Regularity requirements (e.g., Nuclear Regulatory Commission) demand the implementation of test standards, design codes and data bases for CMCs as part of the licensing and approval of this new power plants. Significant progress on standards, codes and databases for heat engine and aero applications has been made since 1991, however new requirements (e.g., durability in extreme temperature and irradiation environments) for CMC applications in NGNP require additional efforts. The current (2006) state of standards, design codes and databases for CMCs is reviewed and discussed. Issues related to development, verification and use of new standards, design codes, and database NGNP applications, including international aspects, are presented and discussed

INTRODUCTION

Ceramic matrix composites (CMCs), although a still-emerging advanced material, may be the only material of choice in Next Generation Nuclear Power (NGNP) plants because of their low thermal expansion, resistance to degradation from irradiation and excellent retention of mechanical properties at elevated temperatures under high pressures.[1] (see Figure 1)

Although CMCs possess many of the critical characteristics of advanced ceramic monoliths (e.g., low density, high stiffness, elevated temperature capability, etc.) they have the advantage of exhibiting increased "toughness" over their monolithic counterparts. [2-3] Continuous fiber-reinforced CMCs exhibit greatly increased "toughness" (i.e., nonlinear energy dissipation during deformation) and therefore provide the inherent damage tolerance, volume/surface area independent properties and attendant increased reliability that are critical in many engineering applications where the brittleness of conventional advanced ceramics makes these materials acceptable.[1]

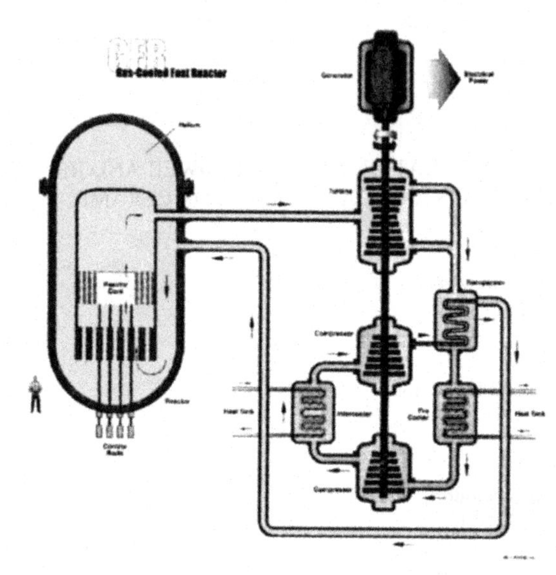

Figure 1 CMC tubes as control rod guides in advanced (GenIV) gas cooled fast reactors

The unique performance of CMCs makes accepted conventional testing and design methods inappropriate. This situation comes just as the characterization and prediction of the thermo-mechanical behavior of CMCs have become the subject of extensive investigations worldwide. The marketplace emergence of CMC prototype/trial products coupled with the relative scarcity of standards (e.g., test methods, practices, guides, classification/terminology, and reference materials) for CMCs and the lack of design codes/data bases for CMCs are limiting factors in commercial diffusion, industrial acceptance and regularity approval[4] of these advanced materials.

In this paper, the issue of standards for CMCs in Gen IV NGNP applications and their extension to design codes and databases is addressed. Standards for CMCs are first defined. This is followed by a discussion of the current status of standards and design codes for CMCs. Finally, some conclusions on the future of standards and codes for CMCs are presented.

STANDARDS

"Standards" as a technical term may have many interpretations[4]. For those in the technical community (e.g., researcher) "standards" may be fundamental test methods or even units of measure. For the end-product user or manufacturer "standards" may be specifications for materials or even requirements for quality control. Commercial standards are equivalent to the rules/terms of information transfer between manufacturers, designers, and product users.[4] Another difference exists between levels of standards: company (internal use and internal consensus); industry (trade use and limited consensus); government (wide use and various levels of consensus); full-consensus (broadest use and greatest consensus).

At this time, there are relatively few (national or international) consensus standards[5] for advanced ceramics and relatively fewer still for CMCs, in particular (see Table 1). For full consensus standards, [i.e., American Society for Testing and Materials (ASTM) Subcommittee C28.07 on Ceramic Matrix Composites, Comité Européen de Normalisation (CEN) Subcommittee TC184/SC1 on Ceramic Composites, and International Organization for Standardization (ISO) Technical Committee TC206 on Fine (Advanced, Technical) Ceramics] various technical and pragmatic issues related to CMC standardization efforts have been presented[5]. Other standards for CMCs, although not full consensus have been introduced through NASA High Speed Research/Enabling Propulsion Program (HSR/EPM) in the United States and Petroleum Energy Center (PEC) in Japan.

Despite these diverse efforts, the paucity of standards has limited the ability to evaluate CMCs on a common-denominator basis and furthermore may be hindering continued development of these advanced materials[4]. Fortunately, although the total number of standards for CMCs is still relatively low, the rate of standards development for CMCs has been increasing. Perhaps more importantly, new efforts within ISO TC206 on Fine Ceramics are aimed at harmonizing existing standards rather than initiating new efforts, thus assisting the introduction of CMC standards.

HARMONIZATION OF STANDARDS

Harmonization of existing standards may play in important role in widespread acceptance and usage of standards for CMCs. An example of such harmonization exists for uniaxial tensile testing which is the most fundamental and, therefore, most common test for CMCs. Currently there are five tensile test method standards for CMCs at room temperature (see Table 1). Of these five standards, three (ASTM, CEN, and EPM) were developed independently. Two others (ISO and PEC) were developed by harmonizing (i.e., choosing the "best" aspects) of the pre-existing standards.

Once aspect of this harmonization is that common terminology allows communication between testers without confusion. In ISO 15733 "Fine ceramics (Advanced ceramics, Advanced technical ceramics)-Test method for tensile stress-strain behaviour of continuous fibre-reinforced composites at room temperature" harmonization has lead to such common terms as fine ceramic and ceramic matrix composite that are defined as follows:

fine (advanced, technical) ceramic, n - A highly engineered, high performance predominately non-metallic, inorganic, ceramic material having specific functional attributes.

ceramic matrix composite, n - A material consisting of two or more materials (insoluble in one another), in which the major, continuous component (matrix component) is a ceramic, while the secondary component/s (reinforcing component) may be ceramic, glass-ceramic, glass, metal or organic in nature. These components are combined on a macroscale to form a useful engineering material possessing certain properties or behavior not possessed by the individual constituents.

In addition to harmonization, ISO 15733 also serves to show the diversity of test methods. For example, no single tensile test specimen geometry has been identified as the "best." As a result, the range of successful test specimen geometries is illustrated in the standard to indicate to users a variety of possibilities (see Figure 2).

Table 1 Selected Test Standards for Ceramic Matrix Composites

Requirement	ASTM	CEN	ISO	EPM (HSR/EPM)	PEC
Bulk CMC, R.T. in Air					
Tension .... In-plane .... Transverse	C1275 C1468	EN658-1 —	ISO15733 —	D-001-93 —	TS-CMC01 —
Compression .... In-plane	C1358	ENV658-2	DIS20504		
Shear .... In-plane .... Interlaminar	C1292 C1292	ENV12289 ENV658-4	DIS20506 DIS20505		— TS-CMC06
Flexure .... Bending .... Shear	C1341 —	ENV658-3 ENV658-5		D-003-93 —	TS-CMC04 —
Mechanical Fatigue ..... Tensile .... Flexural	C1360 —			D-002-93 —	TS-CMC10 —
Density		ENV 1389			
Elastic Constants	C1258				TS-CMC13
Fracture		ENV13234			TS-CMC08 TS-CMC09
Bulk CMC, H.T. in Air/Inert					
Tension .... In-plane .... Transverse	C1359 —	ENV1892 ENV1893		D-001-93 —	TS-CMC01 —
Compression .... In-plane		ENV12290 ENV12291			
Shear .... In-plane .... Interlaminar	— C1425	ENV1894			— TS-CMC06
Flexure .... Bending .... Shear	C1341 —			D-003-93 —	TS-CMC04 —
Mechanical Fatigue ..... Tensile .... Flexural				D-002-93 —	TS-CMC10 —
Thermal Fatigue					TS-CMC14
Creep	C1337	ENV13235		D-004-93	TS-CMC11
Stress Rupture				D-004-93	
Specific Heat		ENV1159-3			
Thermal Diffusivity		ENV1159-2			
CTE		ENV1159-1			
Ceramic Fiber, R.T. in Air					
Tension	C1557	ENV1007-4			
Diameter		ENV1007-1 ENV1007-3			
Density		ENV1007-2			

ASTM =American Society for Testing and Materials. CEN = Comité Européen de Normalisation, ISO = International Organization for Standardization, EPM = Enabling Propulsion Materials. PEC = Petroleum Energy Center

According to ISO1573, a wide range of properties can be extracted from a single uniaxial tensile test of a CMC as illustrated in Figure 3. Each of these properties (e.g., elastic modulus, proportional limit stress, ultimate tensile strength, modulus of toughness) has explicit formulae for determining it from the stress strain curve.

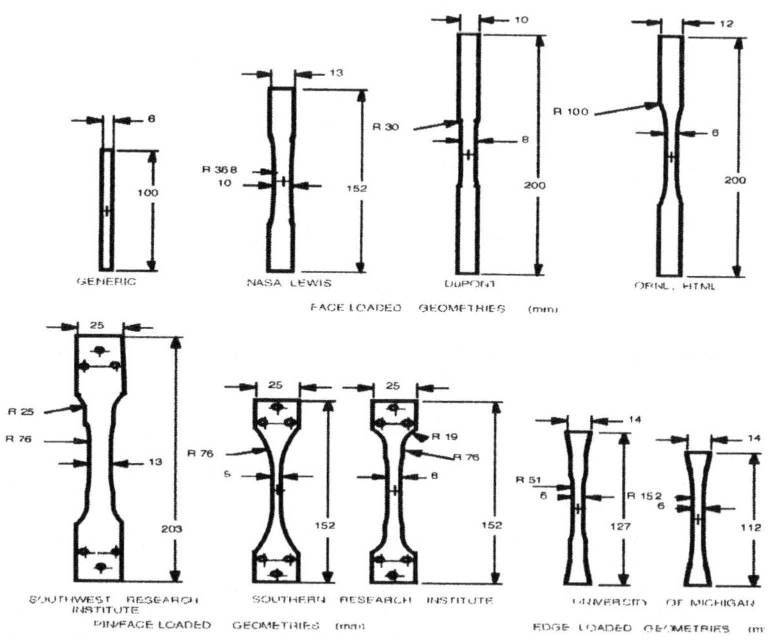

Figure 2 Examples of various tensile test piece geometries contained in ISO15733

For Gen IV gas cooled fast reactor (GFR) applications, a targeted application of CMCs is tubes for control rods. These CMC would be subjected to temperatures of 490 to 850°C at pressures up to 9 MPa of gaseous helium in addition to fast neutron fluence. Standards for mechanical behaviour of CMC tubular components have been addressed only in passing discussions. Standards for irradiation effects of CMCs have not even been proposed.

DESIGN CODES

"Design code" as used in this discussion is not a design manual (i.e., a "cookbook" design procedure that leads to a desired component or system). Instead, design codes are broadly-accepted, general rules for the fabrication of components or systems. A primary objective is the relatively-long safe-life of the design while providing for the reasonably certain protection of life and property. Even though the safety of the design can never be compromised. the needs of the users, manufacturers and inspectors are recognized.

"Design codes" allow flexibility for new designs that are required for performance, efficiency. usability, or manufacturabilty while still providing constraints for safety. Such a wide ranging "design code" incorporates links between materials, general design (formulae, loads, allowable stress, permitted details), fabrication methods, inspection, testing, certification, and data reports, and finally quality control to insure that the "design code" has been followed

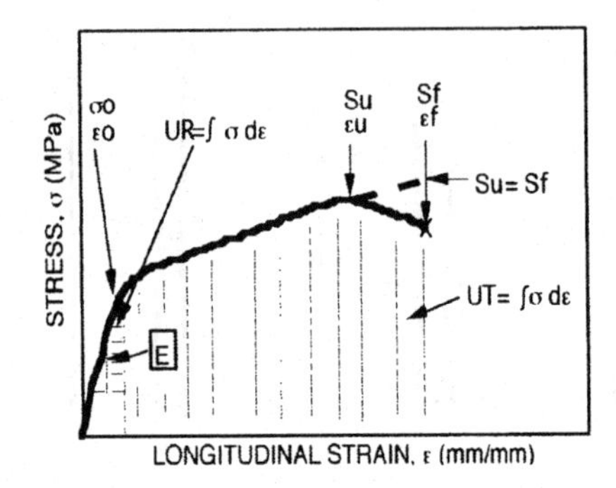

Figure 3 Some explicit parameters extractable from stress-strain curves (ISO15733)

Many of the standards for materials testing, characterization, and quality control are implicit in the "design code". Finally, unlike standards that do not provide for compliance or accountability, "design codes" require compliance through documentation, and certification through inspection and quality control.

There are currently two efforts underway to establish design codes for CMCs: ASME Task Group on Ceramic and Graphite Pressure Equipment and Mil-Hdbk-17 CMC effort. Standards are incorporated in both these efforts to provide consensus methods for determining the properties and performance of materials contained in the "design code". Figure 4 is an illustration of the use of "design by analysis" in which the long-term performance of a component is predicted using an algorithm that requires information on long-term performance of materials determined using standards.

"Design codes" (and imbedded standards) as approved and embraced by industry are particularly important for GenIV nuclear power applications because of the oversight by the U.S. Nuclear Regulatory Agency (NRC). The NRC is the "watchdog" agency of the nuclear power sector and as such is insistent on documentation, best practices and rigorous procedures. Design codes along with their imbedded standards represent debated, balloted and approved consensus documents which are crucial to meeting the approval of the NRC.

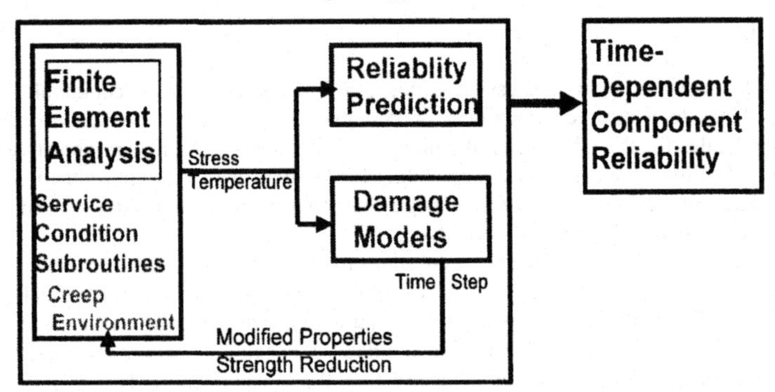

Figure 4 Design by analysis approach for CMC "design code"

CONCLUSIONS

Although the number of standards for CMCs is now over 60 since 1991, these standards have numerous uses including harmonizing international standards, and establishing the basis for determining properties and performance for "qualified materials" within design codes.

Unfortunately, despite this level of development of standards, codes and data bases for CMCs, the current state of standards and codes for CMC applications in GenIV nuclear power generation is inadequate. New standards need to be developed for non-standard tubular shapes under a variety of loading (e.g., tension, flexure, internal pressurization) and environmental (e.g., elevated temperature, high pressure gas and fast neutron fluence). In addition, new design codes must be developed to integrate CMCs into evolving GenIV NGNP applications. Finally, these new design codes must be approved by the U.S. NRC before the use of CMCs in GenIV NGNP applications can be ruled successful.

REFERENCES

[1] U.S. DOE Nuclear Energy Research Advisory Committee, "A Technology Roadmap for Generation IV Nuclear Energy Systems," GIF-002-00, U.S. Department of Energy, Washington, DC (2002)

[2] Jenkins, M. G., Piccola, J. P. Jr., Mello, M. D., Lara-Curzio, E., Wereszczak, A. A. (1993) "Mechanical Behaviour of a 3-D Braided, Continuous SiC Fiber-Reinforced / CVI SiC Matrix Composite at Ambient and Elevated Temperatures," *Ceramic Engineering and Science Proceedings*, **14** , 9-10. 991-997 (1993)

[3] Karnitz, M. A., Craig, D. A. and Richlen, S. L. (1991) "Continuous Fiber Ceramic Composite Program," *Ceramic Bulletin*, **70**, 3, 430-435 (1991).

[4]Schneider, S. J. and Bradley (1988) "The Standardization of Advanced Ceramics," Advanced Ceramic Materials, Vol 3, No. 5, pp. 442-449.

[5]Jenkins, M. G., "Standards for CMCs: What are They Good for?" HT-CMC5 Proceedings, American Ceramic Society, Westerville, OH (2005)

DETERMINATION OF PROMISING INERT MATRIX FUEL COMPOUNDS

C.R. Stanek, J.A. Valdez, K.E. Sickafus, K.J. McClellan

MST-8, Structure and Property Relations
Los Alamos National Laboratory
Los Alamos, NM, 87545

R.W. Grimes
Dept. of Materials
Imperial College London
London, UK, SW7 2BP

ABSTRACT

The stockpile of civil (and weapons grade) plutonium could be reduced if it were incorporated into an inert matrix fuel (IMF) for insertion in a power reactor. However, potential IMF compounds must meet a series of requirements including acceptable thermal conductivity and suitable radiation tolerance. Atomistic simulations are employed to investigate properties related to IMF applicability in oxide materials $RE_2Ce_2O_7$ (where RE = a rare earth atom or Y) pyrochlores. In particular, radiation tolerance is predicted through these simulations. The theoretical results are then compared to experimental irradiation data in order to determine the applicability of these compounds for further consideration as potential inert matrix fuels.

INTRODUCTION

In order to increase nuclear energy production, the issue of how to manage the nuclear waste produced must be addressed. Furthermore, it is well known that security risks exist as a direct result of the world's plutonium stocks[1]. To address these concerns, the Advanced Fuel Cycle Initiative (AFCI) is developing a technology base to demonstrate the practicality of advanced fuel cycles. Inert matrix fuels (IMFs) are being considered for use in three distinct fuel cycle concepts[2], namely: (1) burning of Pu in existing light water reactors (LWRs), (2) Pu burning in fast reactors and (3) the transmutation of minor actinides (MAs). Since the latter two scenarios involve the construction of modern facilities while the first scenario makes use of existing reactors, these concepts can be thought of as ordered chronologically in terms of potential implementation.

Simply, an inert matrix fuel is a material that burns Pu and other actinides without breeding Pu (or other actinides). Therefore, IMFs can be used to produce energy while eliminating actinide excesses (of which Am contributes 10% of the radiotoxic inventory). "Non-fertile" IMFs do not contain uranium and therefore produce no fissionable isotopes while "low-fertile" IMFs contain uranium and therefore breed Pu and other MAs.

Table I describes the important materials properties for several potential inert matrix fuel compounds. Each of the properties included in Table I must be accounted for when designing an IMF, though the desired value of the property will vary according to the application. For example, c-ZrO_2 is difficult to reprocess. This makes it a desirable candidate for a "once through" IMF (*i.e.* fuel not recycled after use), as the difficulty in reprocessing increases proliferation resistance. However, this property clearly hinders the use of c-ZrO_2 in an advanced fuel cycle where the fuel is reprocessed.

	c-ZrO_2	MgO	$MgAl_2O_4$	$A_2B_2O_7$	NiAl	RuAl	TiN	ZrN
Melting point (°C)	2675	2832	2135	~2200	1638	2050	2930	2980
Thermal conductivity (W/mK)	~4	~10	<5.9	1.5 - 2.5	80	50 - 70	~19	~11.0 - 24.0
Thermal expansion (x 10^{-6}/°C)	7.5 - 13	~13.5	~7.5	~8.0 - 11.0	~12.0 - 15.0	5.5 - 11	9.4	6.5 - 11
Heat capacity (@1000 °C, J/mol K)	78	317	190	~280	<60		55	55
Fabricability	standard cold press sinter technique				variety of methods		sintering aids or hot press	
Reprocessability	poor	good	poor	poor	good		reasonable	
Radiation tolerance	good				good for heavy ion		good (SP100, heavy ion, etc)	

Table I Values of several materials properties important for inert matrix fuels for potential compounds.

In this paper, we demonstrate how a combination of theoretical and experimental techniques can be employed to aid in the determination of promising inert matrix fuel compounds. Specifically, this work addresses the radiation tolerance of $RE_2Ce_2O_7$ (where RE = a rare earth atom or Y) pyrochlores. The goal of this work is to provide information that can be used to stragetically choose compounds for further consideration in test reactors.

INTEGRATION OF ATOMISTIC SIMULATION AND MODELING

Although irradiating potential inert matrix fuel compounds in reactor conditions provides essential information pertaining to the validity of the material in actual conditions, these time consuming and expensive experiments can be strategically designed based upon results of the combination of atomistic simulation and ion beam irradiation. In particular, radiation damage has been successfully examined via this combination of techniques. The atomic scale simulation component of this technique considers a wide range of materials in order to highlight regions of compositional interest. Compounds can then be chosen strategically and probed further with ion beam irradiations and subsequent electron microscopy. Employing both techniques allows for radiation tolerance trends to be established expeditiously for a wide range of compounds.

The atomistic simulations described here were based upon a Born-like description of the lattice[3], using Buckingham potentials[4] to describe the short-range interaction between ions. The lattice energy is therefore:

$$E_{lattice} = \frac{1}{2}\sum_{i}\sum_{j\neq i}\left[\frac{q_i q_j}{4\pi\epsilon_0 r_{ij}} + A_{ij}\exp\left(-\frac{r_{ij}}{\rho_{ij}}\right) - \frac{C_{ij}}{r_{ij}^6}\right] \quad (1)$$

where q is the charge of the ion, ε_0 is the permittivity of free space, r_{ij} is the interionic separation, and A_{ij}, ρ_{ij} and C_{ij} are the adjustable potential parameters for each ion i and j. The potentials used in this work can be found elsewhere[5-6], though it is important to note that all potentials employed were derived self-consistently. The polarizability of ions is accounted for by the shell model[7].

In this work, we have employed these atomistic simulation techniques to model the disorder (and therefore, radiation tolerance) of a range of $A_2B_2O_7$ pyrochlore oxides (where A is a 3+ rare-earth cation ranging in ionic radius from Lu^{3+} to La^{3+} and B is a 4+ cation ranging from Ti^{4+} to Pb^{4+}). To visualize the crystal structure of pyrochlore, we have used the convenient, fluorite-type description[8], see Figure 1. The pyrochlore structure can thereby be considered as an ordered, defective fluorite solid solution. In CaF_2 (the mineral "fluorite"), the fluorine anions are located in the tetrahedral sites of a Ca face centered cubic array. In this description of pyrochlore, the A and B cations form the face centered cubic array, but are additionally ordered in the <110> direction such that the A cations are eight coordinate and the B cations are six coordinate with respect to oxygen. This cation ordering means that the tetrahedral anion sites are no longer crystallographically identical. In fact, there are now three distinct tetrahedral sites: the 48*f*, which has two A and two B nearest neighbors, the 8*a*, which has four B nearest neighbors and the 8*b*, which has four A nearest neighbors. In pyrochlore, the 8*a* positions are vacant. Figure 1 depicts one eighth of the pyrochlore unit cell, which is, as can be seen, analogous to a single fluorite unit cell.

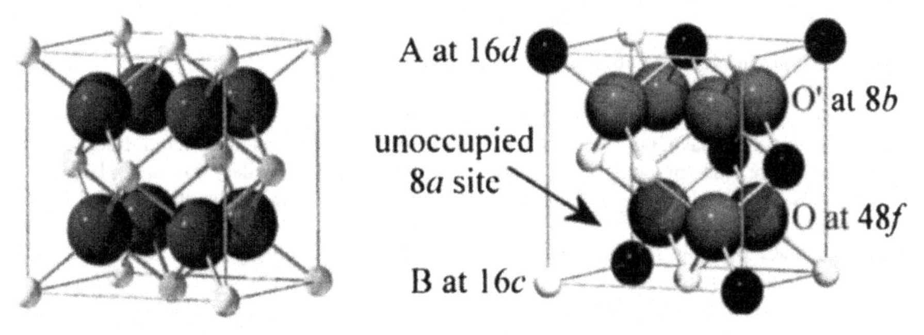

Figure 1 Comparison between a fluorite unit cell (left) and 1/8 of a pyrochlore unit cell (right).

Previous work has determined the disorder process in pyrochlores as a combination of cation antisite and anion Frenkel disorder[5,9-10], which can be expressed in Kröger-Vink notation as:

$$A_A^x + B_B^x + O_O^x \rightarrow \{A'_B + B^{\cdot}_A + V^{\cdot\cdot}_{O(48f)} + O''_{i(8a)}\}^x \quad (2)$$

An explanation for this process is that cation disorder disrupts the coordination of the crystallographically unique oxygen lattice sites, thereby leaving them more similar to a fluorite environment. Therefore, cation antisite disorder effectively lowers the energy for anion Frenkel disorder. Furthermore, this type of disorder can be expected to be induced readily by radiation. Figure 2 depicts the results of calculations describing the energy of the above reaction, and therefore the propensity for a material to accommodate the types of defects associated with radiation damage. The results from these calculations are provided in the form of a contour plot (see Figure 2), where $A_2B_2O_7$ compounds are ordered by their cationic radii: increasing A radius along the ordinate and B along the abscissa. Thus each compound occupies a point on the map. Black points refer

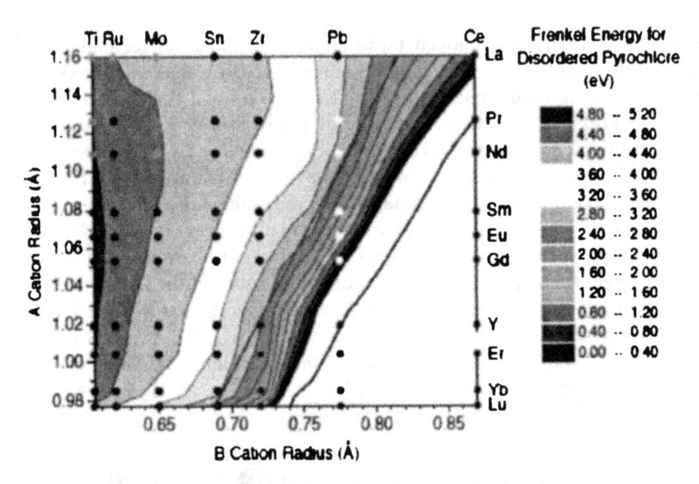

Figure 2 Contour plot of calculated pyrochlore disorder energies.

to a stable pyrochlore former for which calculations were carried out. Red points refer to compounds which have not been observed to form, *i.e.* exist as disordered fluorites (thus showing where the pyrochlore to fluorite compositional boundary occurs). Here, such compounds were modeled in the pyrochlore structure since this facilitates the construction of a continuous pyrochlore contour map. For each compound modeled, a defect energy process is calculated and plotting software generates contours of equal energy over the cation radius surface. Cationic radii and reaction energies are converted to a matrix via the Kriging method (using a smoothness value of 0.5). This method is an optimized linear interpolation from which 3D surfaces can be generated. Using this approach, areas of compositional interest (*i.e.* that have similar process energies as predicted by our methodology) are easily identified. An advantage of this approach is that predictions of properties can also be made for pyrochlore structures that were not explicitly calculated (e.g. here for hafnates: $A_2Hf_2O_7$; Hf ionic radius 0.71Å[11]). Brisse and Knop[12] have shown experimentally that the pyrochlore lattice parameter of lanthanide stannates $Ln_2Sn_2O_7$ varies smoothly as a function of lanthanide (Ln) ionic radius. Furthermore, Kennedy[13] has shown using neutron diffraction that the oxygen positional parameter of the lanthanide stannate pyrochlores also changes smoothly as a function of lanthanide radius. These types of experiments support the use of contour maps based on cationic radii as an ordering parameter. In Figure 2, red areas denote a high-energy process where blue areas correspond to a low-energy process.

As previously mentioned, for $A_2B_2O_7$ pyrochlores, the disorder reaction (*i.e* Equation 2) is directly related to radiation tolerance. Therefore, it is clear from Figure 2 that the $A_2Ti_2O_7$ series of compounds is predicted to be less radiation tolerant than, for example, the $A_2Zr_2O_7$ series of compounds. Indeed, this prediction has been verified experimentally through ion beam implantation and electron microscopy[14].

The predictive capabilities of Figure 2 were previously verified for pyrochlores $Er_2Ti_2O_7$ and $Er_2Zr_2O_7$ (*i.e.* $Er_2Ti_2O_7$ was found to amorphize at radiation doses where $Er_2Zr_2O_7$ remained crystalline). In this work, the verification of the radiation tolerance trend established in Figure 2 is furthered by experimentally considering $RE_2Ce_2O_7$ compounds. Figure 3 depicts the results (in the form of TEM cross sectional images) of ion irradiation experiments conducted on $La_2Ce_2O_7$ and $Y_2Ce_2O_7$. It is clear from Figure 3 that both $La_2Ce_2O_7$ and $Y_2Ce_2O_7$ remain crystalline at high radiation doses. If radiation tolerance is directly related to the energy required to form the defects associated with pyrochlore disorder (i.e. the less energy required to form the defects associated with radiation damage, the greater the radiation tolerance), then the trend observed in Figure 2 clearly predicts that $La_2Ce_2O_7$ and $Y_2Ce_2O_7$ should be more radiation tolerant than the analogous $A_2Ti_2O_7$ and $A_2Zr_2O_7$ compounds.

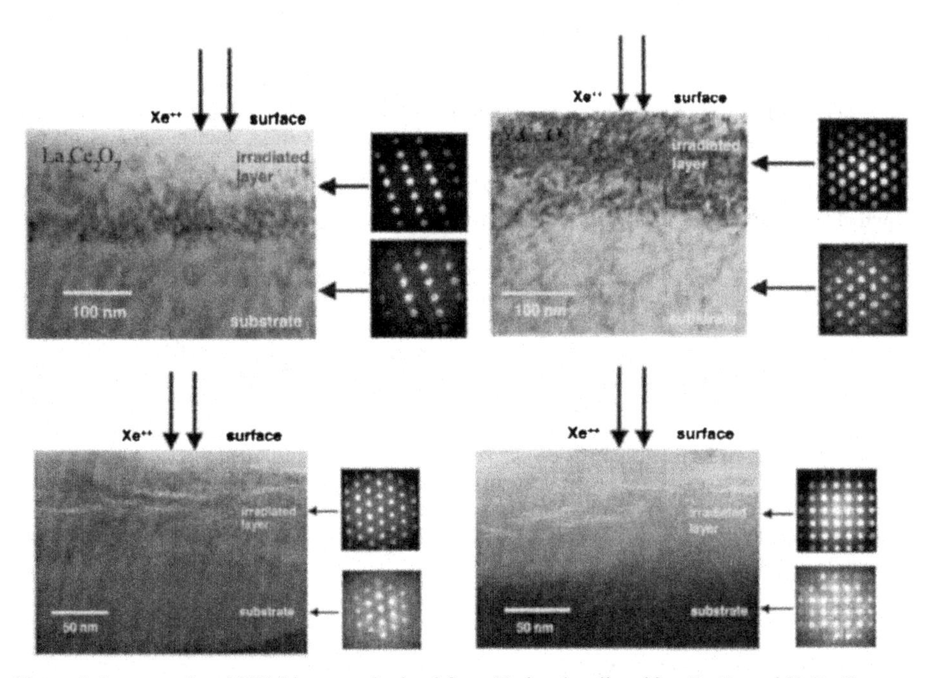

Figure 3 Cross-sectional TEM images obtained from Xe ion-irradiated $La_2Ce_2O_7$ and $Y_2Ce_2O_7$ to fluences of $1x10^{20}$ Xe/cm^2 (top images) and $5x10^{20}$ Xe/cm^2 (bottom images), clearly showing that no amorphization has taken place.

There are a few remaining points that should be made concerning Figure 3. Firstly, both $La_2Ce_2O_7$ and $Y_2Ce_2O_7$ are not stable pyrochlore formers. Rather, these compounds form as disordered fluorites. It therefore follows that defects associated with radiation damage occur spontaneously in these compounds. In fact, the white area in Figure 2 (the disorder energy value for $Y_2Ce_2O_7$) effectively corresponds to a negative disorder energy, suggesting that the reaction in Equation 2 spontaneously occurs. Furthermore, a rather straightforward implementation of an inert matrix fuel would be with a compound that resembled the current fuel type, which is predominantly UO_2. Therefore, that $La_2Ce_2O_7$ and $Y_2Ce_2O_7$ do not crystallize in the pyrochlore structure does not necessarily prevent them from being candidates for inert matrix fuels. In fact, there is no inherent advantage corresponding to the pyrochlore structure that makes it more suitable for this application than fluorite structured materials (or other structures for that matter).

A second worthwhile point is that Ce is a common surrogate for Pu. For example the 4+, VI coordinate ionic radii differ by only 0.01Å[11] and the BO_2 lattice parameters differ by only 0.0142Å. (It should be noted that $A_2B_2O_7$ pyrochlore formation occurs via the reaction: $A_2O_3 + 2MeO_2 \rightarrow A_2B_2O_7$). Therefore, these compounds represent potential Pu containing pyrochlore IMFs.

CONCLUSIONS and FUTURE WORK

We have used a combination of theoretical and experimental techniques in order to determine radiation tolerance behaviour of potential IMFs. This information can be used to determine which compounds are worth investigating further, while simultaneously isolating compounds that are not likely to be viable fuels. There is practical evidence for the success of these studies. Preliminary plans for the LWR-2 irradiation at the Advanced Test Reactor include pyrochlore inert matrix fuels. The results of this work are especially valuable when combined with post irradiation examinations.

The work described here only provides information for the radiation tolerance of potential inert matrix fuels. Another important property that can be probed with these techniques is thermal conductivity. The thermal conductivity of pyrochlores has been recently predicted by, Schelling, *et al.*[15]. The trends generated by the simulation can be subsequently verified by, for example, using the laser flash method. However, on pristine samples, these results only describe the thermal conductivity before irradiation. In pile, it can be expected that the thermal properties of the fuel will vary. Therefore, it is desirable to determine the thermal conductivity of inert matrix fuels as a function of radiation damage. For example, materials that are not currently considered as inert matrix fuels on account of undesirable thermal conductivity may become viable as their thermal conductivity does not decrease as readily as materials with a high initial thermal conductivity. Both simulations and experiments can be performed to determine this property variation as a function of radiation damage.

REFERENCES

[1]National Academy of Sciences, *Management and Disposition of Excess Weapons Plutonium* (1994).

[2]R. Chawla and R.J.M. Konings, "Categorisation and Priorites for Future Research on Inert Matrix Fuels: An extended synthesis on the panel discussions," *Prog. Nucl. Energy*, **38**, 455-458 (2001).

[3]M. Born, *Atomtheories des Festen Zustandes*. Teubner, Leipzig, Germany, 1923.

[4]R.A. Buckingham, "Classification of State Gaseous He, Ne and Ar," *Proc. Roy. Soc. London*, **168A**, 264-283 (1938).

[5]L. Minervini, R.W. Grimes and K.E. Sickafus, "Disorder in Pyrochlore Oxides," *J. Am. Ceram. Soc.*, **83**, 1873-78 (2000).

[6]C.R. Stanek, L. Minervini and R.W. Grimes, "Nonstoichiometry in $A_2B_2O_7$ Pyrochlores," *J. Am. Ceram. Soc.*, **85**, 2792-98 (2002).

[7]B.G. Dick and A.W. Overhauser, "Theory of Dielectric Constants of Alkali Halide Crystals," *Phys. Rev.*, **112**, 90-103 (1958).

[8]E. Aleshin and R. Roy, "Crystal Chemistry of Pyrochlore," *J. Am. Ceram. Soc.*, **45**, 18-25 (1962).

[9]P.J. Wilde and C.R.A. Catlow, "Defects and Diffusion in Pyrochlore-Structured Oxides," *Solid State Ionics*, **112**, 173-183 (1998).

[10]P.J. Wilde and C.R.A. Catlow, "Molecular Dynamics Study of the Effect of Doping and Disorder on Diffusion in Gadolinium Zirconate," *Solid State Ionics*, **112**, 185-195 (1998).

[11]R.D. Shannon, "Revised Effective Ionic Radii and Systematic Studies of Interatomic Distances in Halides and Chalcogenides," *Acta Cryst. A*, **32**, 751-67 (1976).

[12]F. Brisse and O. Knop, "Pyrochlores III. X-ray, Neutron, Infrared and Dielectric Studies of $A_2Sn_2O_7$ Stannates," *Can. J. Chem.*, **46**, 859-73 (1968).

[13]B.J. Kennedy, "Structural Trends in Pyrochlore-type Oxides," *Physica B*, **241**, 303-10 (1998).

[14]K.E. Sickafus, L. Minervini, R.W. Grimes, J.A. Valdez, M. Ishimaru, F. Li, K.J. McClellan and T. Hartmann, "Radiation Tolerance of Complex Oxides," *Science*, **289**, 748-51 (2000).

[15]P.K. Schelling, S.R. Phillpot and R.W. Grimes, "Optimum Pyrochlore Compositions for Low Thermal Conductivity," *Phil. Mag. Lett.*, **84**, 127-137 (2004).

DENSIFICATION MECHANISM AND MICROSTRUCTURAL EVOLUTION OF SiC MATRIX IN NITE PROCESS

Kazuya Shimoda
Graduate School of Energy Science, Kyoto University
Gokasho Uji, Kyoto 611-0011, Japan

Joon-Soon Park, Tatsuya Hinoki and Akira Kohyama
Institute of Advanced Energy, Kyoto University
Gokasho Uji, Kyoto 611-0011, Japan

ABSTRACT

Monolithic SiC and unidirectional SiC/SiC composites were prepared by NITE process, using SiC "nano"-slurry infiltration technique, and the effects of tailoring raw materials and of processing conditions on density, microstructural evolution and mechanical properties were investigated. In particular, the mechanism on SiC matrix densification was discussed by the characterization using monolithic SiC fabricated by NITE process at the same condition for SiC/SiC composite fabrication. The densification was dramatically promoted at the temperature of 1750-1800°C and saturated above 1800°C. Use of nano-sized powder is effective to increase capillary adhere force, resulting in denser matrix below 1900□ than submicron-powder. Additional SiO_2 enables low temperature and viscosity of liquid phase, and thus increases capillary adhere force. Furthermore, additional SiO_2 content plays significant roles to control precisely gas generation of volatile species. SiC grain growth can be controlled by nano-sized powder characteristics and process additives as well as processing conditions. Matrix densification mechanism was proposed for the case of NITE process based on the models of the following three parameters: (1) capillary adhesion force, (2) gas generation and (3) grain growth. The conditions, based on the dense matrix of monolithic SiC, were applied to SiC/SiC composites though NITE process.

INTRODUCTION

Advanced nuclear energy systems, such as gas cooled fast reactor (GFR), very high temperature reactor (VHTR) and fusion reactor are potential candidates for sustainable energy systems in the future. In order to realize these attractive energy systems, structural materials must be responsible to keep their performance under very severe environment including high-temperature, high energy neutron bombardment and surrounding coolants and fuels.

Continuous silicon carbide (SiC) fiber reinforced SiC matrix (SiC/SiC) composites have been recognized as attractive candidate structural materials and components in these advanced energy systems due to their potentiality for providing excellent mechanical properties at high-temperature and low induced radioactivity [1-4].

Recent progress in qualification of the reference SiC/SiC composites for these systems is rationalizing the innovative processes which potentially provide dense matrix with high-crystalline SiC and near-stoichiometry. High-crystalline SiC and near-stoichiometry are most preferred materials, because they possess proven environmental resistance [4-6]. From this aspect, technology based on chemical vapor infiltration (CVI) is the most mature and is used worldwide [7-9]. However, this process requires very long manufacturing time to densify thick cross-sections. In additions, it becomes very expensive and difficult to fabricate large size and complex shape composites. Dense matrix with low porosity is strongly required for high thermal conductivity, high strength and reducing of the leakage helium gas as a coolant gas [10, 11]. The composites, therefore, need to be dense for achieving these performances. However, the SiC/SiC composites fabricated by conventional processes (CVI and polymer impregnation and pyrolysis (PIP) process), in general, have pore volume fraction of 10~20% [7-9, 12, 13].

Recently, a new process called Nano-Infiltration Transient Eutectic-phase (NITE) process, using SiC "nano"-slurry infiltration technique, has been developed which produces nearly-full dense matrix with well-crystallized SiC at the "lower" sintering temperature (above 1800°C) while protecting SiC fibers and interphase after induced processing conditions in our group at Kyoto University[14-16]. Very limited information has been obtained about the mechanism on matrix densification. For the establishment of NITE process toward a large scale fabrication with sufficient margins in process condition, the characterization of SiC powder and the contributions of processing additives, which play important roles to form dense matrix, were studied using monolithic SiC fabricated by NITE process at the same condition for SiC/SiC composite fabrication. Effects of process conditions (sintering temperature/holding time) on density, microstructure and mechanical properties of NITE-SiC and -SiC/SiC composites were characterized as another key.

EXPERIMENTAL PROCEDURE

Monolithic SiC fabrication through NITE process

To investigate the effects of SiC particle size on the densification, two kinds of commercially available beta-SiC powder: submicron-sized SiC powder (Ibiden Corp., Japan) and nano-sized SiC powder (Marketech International Inc., USA) were prepared as the starting powder. The morphology of SiC powders were investigated by field emission scanning electron microscopy (FE-SEM), X-ray diffractmetry (XRD), X-ray Photoelectron Spectroscopy (XPS). The average particle size of beta-SiC powders were 20 nm and 0.32 um, respectively, as observed using FE-SEM. Al_2O_3 (Sumitomo Chemical Industries Ltd., Japan), Y_2O_3 (Kojundo Chemical Industries Ltd., Japan) and SiO_2 (Kojundo Chemical Industries Ltd., Japan) were used as processing additives. SiC powders were ball-milled with them for 5 hours to form SiC slurry. The mixed slurry was dried, and then sintered by hot-pressing in Ar atmosphere under 20MPa pressure. Temperature was varied from 1700°C to 1900°C, while holding time was changed from 0 hour to 5 hours. Sintered sample size was cut into 4^W x 22^L x 2^T mm^3. Density of sintered body was determined by the Archimedes principle, using distilled water as the immersion medium. Theoretical density of sintered sample was calculated by following the ratio of a mixture of SiC powder and processing additives. The 3-point flexural test was carried out using the number of 3-5 specimens, with crosshead speed of 0.5 mm/min at room-temperature in an INSTRON 5581 test machine. Flexural stress (σ), flexural strain (ε) and modulus of elasticity in bending (E) were calculated by the following equation (1), (2) and (3), respectively.

$$\sigma = 3PL/2wt^2 \qquad (1)$$

$$\varepsilon = 6Dt/L^2 \qquad (2)$$

$$E = 0.25L^3m/wt^3 \qquad (3)$$

, where P is load at given point in test, L is outer support span, w is specimen width, t is specimen thickness, D is deflection at beam center at a given point in the test and m is slop of tangent to the initial straight-line portion of the load-deflection curve. Fracture after flexural test was observed by using FE-SEM with EDS.

SiC/SiC composites fabrication through NITE process

Tyranno^{TM}-SA grade-3 polycrystalline SiC fibers (Ube Industrials Ltd., Japan) were used as the reinforcement to fabricate SiC/SiC composites. Unidirectional (UD) uncoated and Pyrolytic carbon (PyC)-coated fibers were impregnated in "nano"-slurry, which mixed SiC nano-sized powder and small amount of processing additives. The thickness of pyrolytic carbon (PyC) interphase deposited on the fiber surface through chemical vapor deposition (CVD) was about 0.5um. Those prepared green sheets were stacked unidirectionally in a graphite die, and then hot pressed at 1800°C for 2 hours in Ar atmosphere under a pressure of 20MPa after drying. Fabricated composites were also evaluated using the same experimental procedures as monolithic SiC.

RESULTS AND DISSCUSSIONS

Exploring Fabrication Conditions (monolithic SiC)

· Tailoring raw materials

For SiC/SiC composite fabrication by hot-pressing, potential fiber degradation caused by high temperature and pressure needs to be considered. If the SiC matrix could be sufficient densification, the processing temperature and pressure should be as low as possible. Liquid phase sintering (LPS) technique, which can be densified at "low" temperature (~2000°C) and produce toughed SiC ceramics [17]. Al_2O_3-Y_2O_3 was extensively employed as processing additives. Fig. 1 shows the relative density of nano- vs. submicron-sized SiC powder as the function of sintering temperature. Use of SiC nano-sized powder was effective way to promote the densification at lower temperature (below 1900°C). This is supposed to be because finer particle enhanced densification by the production of desirable conditions, where boundary- and volume-diffusion from boundary were promoted due to the increasing of the number of interparticle contacting points. In particular, capillary adhere force between particles was greatened by use of finer particle in the case of LPS process, because friction between particles is decreased and particles are rapidly reconstructed by liquid phase formed at eutectic-point. To lower processing tempera-

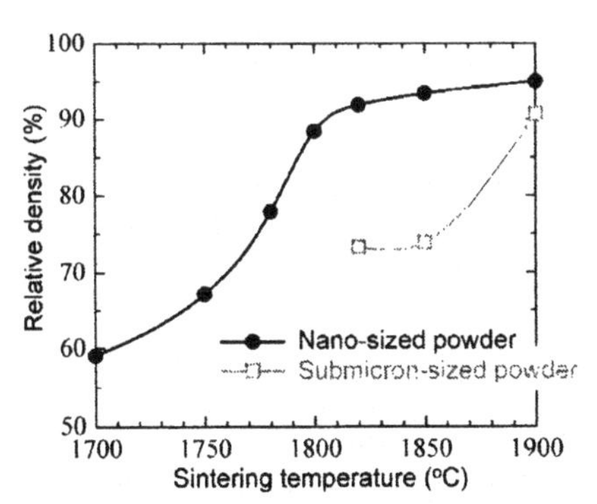

Fig. 1 Relative density of nano- vs. submicron-sized SiC powder as the function of sintering temperature.

Table. 1 Characterization of SiC nano-sized powder.

Crystalline structure	Particle size	Apparent particle density	Particle density	C/Si ratio	Free carbon	Surface chemistry
Beta-SiC(3C)	5-20nm	$0.08Mg \cdot m^{-3}$	$2.94Mg \cdot m^{-3}$	1.26-1.29	2.7wt%	C-C, Si-C

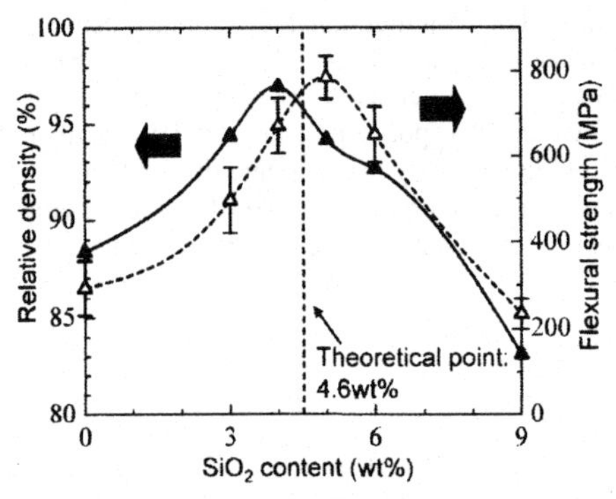

Fig. 2 Relative density and flexural strength as the function of additional SiO_2 content with constant Al_2O_3-Y_2O_3 content of 9wt% at 1800°C.

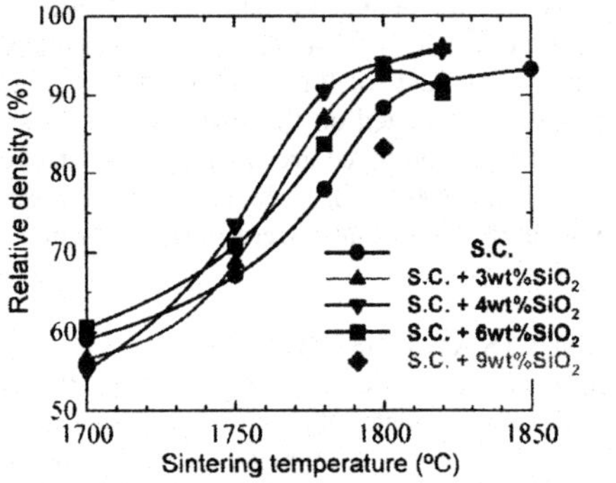

Fig. 3 The effects of additional SiO_2 on the relative density as the function of sintering temperature. Standard condition (S.C.): 9wt%Al_2O_3-Y_2O_3

-ture, SiC nano-sized powder was used at later experiment. We reported that the size of SiC nano-sized powder had a close association with surface structure [18]. Characteristics of SiC nano-sized powder was summarized in Table. 1. SiC nano-sized powder used in this study was typical of being covered with free carbon of the amount of 2.7wt%. Fig. 2 shows the relative density and flexural strength as the function of additional SiO_2 content with constant Al_2O_3-Y_2O_3 content of 9wt% (standard condition: S.C.) at 1800°C. The highest density and strength was achieved at 4wt% of SiO_2 addition. Excess SiO_2 (above 4wt%) caused a significant decrease of bulk density, due to the formation of large amount of pores (see Fig. 4). Following reactions between SiC and SiO_2 are possible to explain these phenomena.

$$SiO_2\ (s, l) + 3C\ (s) = SiC\ (s) + 2CO\ (g) \qquad (1)$$

$$2SiO_2\ (s, l) + SiC\ (s) = 3SiO\ (g) + CO\ (g) \qquad (2)$$

Judging from thermodynamics, reaction (1) dominates at the region of sintering temperature (1800°C). Additional SiO_2 removed the excess carbon near its surface (reaction (1)), resulting in the improvement of densification furthermore. However, adding excess SiO_2 promoted the evolution of volatile species, like CO (g) and SiO (g) via reaction (2). The reaction between SiC and Al_2O_3 as follow could not be contributed to gas evolution at the temperature of 1800°C because the sum of Al_2O, SiO and CO isn't exceed the total pressure (log $P_{total} = 0$), as reported elsewhere [19]. It is also hardly expected of the gas species evolution caused by Y_2O_3.

$$Al_2O_3\ (s) + SiC\ (s) = Al_2O\ (g) + SiO\ (g) + CO\ (g) \qquad (3)$$

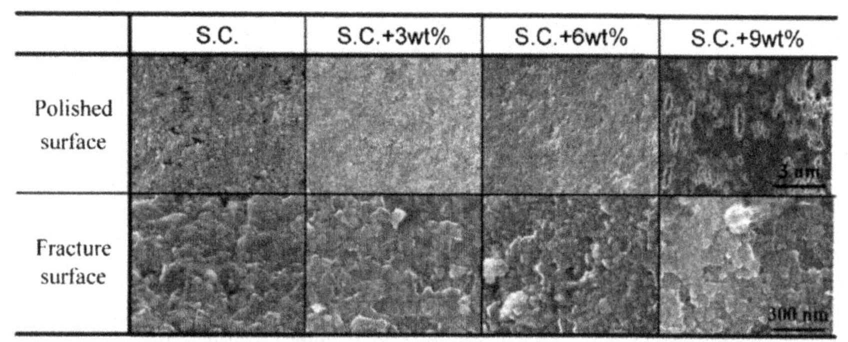

Fig. 4 FE-SEM micrographs of polished and fracture surface of monolithic SiC sintered at 1800°C

Because SiC nano-sized powder contained excess carbon of 2.7wt%, 4.6wt% was theoretically calculated as an amount of additional SiO_2 for only reaction (1). The highest peak of density and strength lied near this theoretical calculation. In the case of SiC nano-sized powder including excess carbon, it was revealed that additional SiO_2 content was able to be quantitatively calculated by the amount of excess carbon.

· Process conditions (sintering temperature & holding time)

Fig. 3 shows the effects of additional SiO_2 on the relative density as the function of sintering temperature. The densification was dramatically promoted at the temperature of 1750-1800°C and saturated above 1800°C. In particular, Additional SiO_2 is effective way on densification of this temperature rage furthermore. This can also been demonstrated by observing microstructural evolution and fracture surface shown in Fig. 4. The homogenous grain growth of fine SiC particles developed from 1750°C in the case of additional SiO_2, resulting in the progress of densification and the decrease of pores on triple points. Some of coarse SiC grains may cause transcrystalline fracture without additional SiO_2. This is possible to explain by following two re-

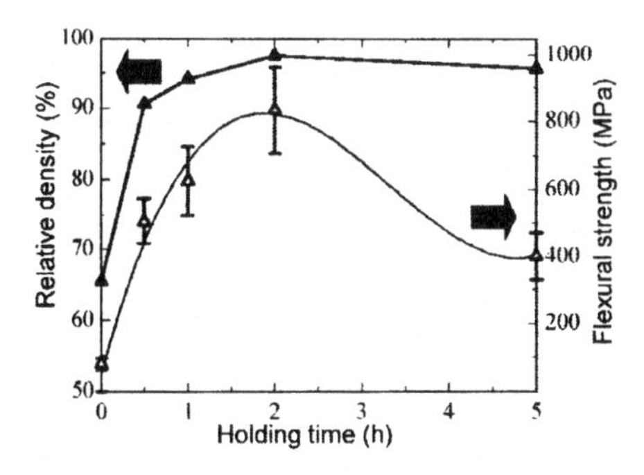

Fig. 5 Relative density and flexural strength of monolithic SiC with S.C. + 3wt%SiO_2 at 1800C as the function of holding time.

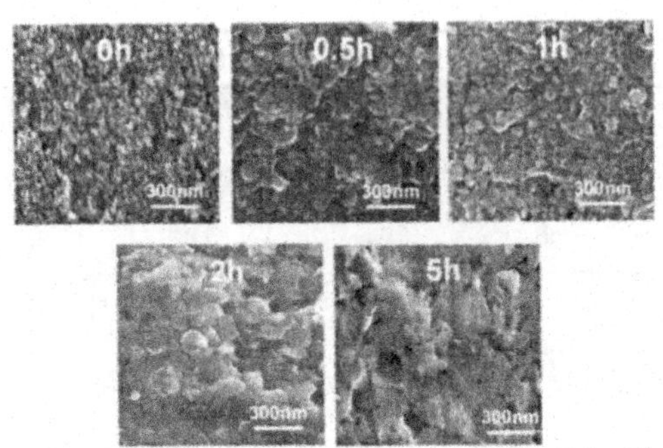

Fig. 6 FE-SEM photographs of fracture surface of monolithic SiC with S.C. + 3wt% SiO_2 at 1800°C under various holing time.

-asons. At first, additional SiO_2 promotes to form liquid phase at lower temperature. Moreover, additional SiO_2 removed the excess carbon on its surface (reaction (1)), resulting in improvement of densification. On the contrast, 9wt% of SiO_2 content decreased the bulk density at 1800°C due to the formation of large amount of pores. This may be due to a lot of small pores formed gas evolution reaction of relating processing additives via reaction (2) and reaction (3).

The holding time was varied from 0 hour to 5 hours in order to improve the densification at 1800°C furthermore using processing additives of lower SiO_2 content (S.C. + 3wt% SiO_2). Fig. 5 shows the effect of holding time at 1800°C on relative density and flexural strength. The densification process was dramatically promoted even by holding for 30 minutes. The near full-density of theoretical density was achieved by holding for 2 hours. Similarly, maximum strength (about 800MPa) was achieved by holding for 2 hours. However, when holding for 5 hours, the density and strength were decreased. This might be due to the extreme growth of SiC grains and a lot of small pores formed gas evolution reaction of relating processing additives via reaction (2) and reaction (3), as shown in Fig. 6. Therefore, the control of SiC grain growth was essential to obtain dense matrix with high strength. It was conclude that optimum holding time was 2 hours in order to obtain excellent densification and strength at 1800°C in the case of lower processing additives (S.C. + 3wt% SiO_2).

Exploring fabrication conditions (SiC/SiC composites)

SiC/SiC composites through NITE process were fabricated under exactly the same optimum conditions as the monolithic SiC. Fig. 7 shows the difference in microstructure of SiC/SiC composites using unidirectional uncoated or pyrolytic carbon (PyC) coated fibers as the reinforcement. Relative density was achieved about 98% of theoretical full density due to sufficient infiltration of SiC nano-sized powder into narrow region within fiber bundle in the case of using uncoated fibers. Moreover, this SiC/SiC composite exhibited excellent average flexural strength (about 1.3GPa). However, the reaction between fiber and matrix progresses extremely, so that the strain-stress curve completely displayed catastrophic fracture behavior, as shown in Fig. 7. Whereas, the SiC/SiC composites using fibers coated by CVD process, whose strain-stress curve showed non-catastrophic fracture behavior, had small decrease on the bulk density with small amount of open pores at inter-fibers, compared with SiC/SiC composites using uncoated fibers. For all that, its density is comparatively higher than that of conventional CVI-, PIP-SiC/SiC composites. As shown in Fig. 8, this may be supposed that macro-pores

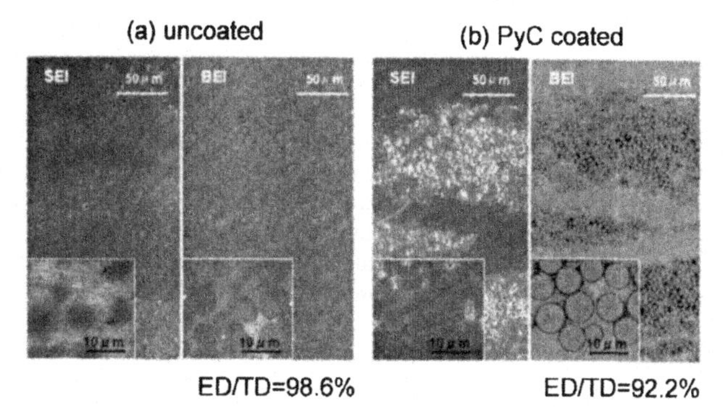

Fig. 7 Polished cross-section of SiC/SiC composites with (a) uncoated and (b) PyC coated fibers. SEI (Secondary Electron Image), BEI (Backscattered Electron Image)

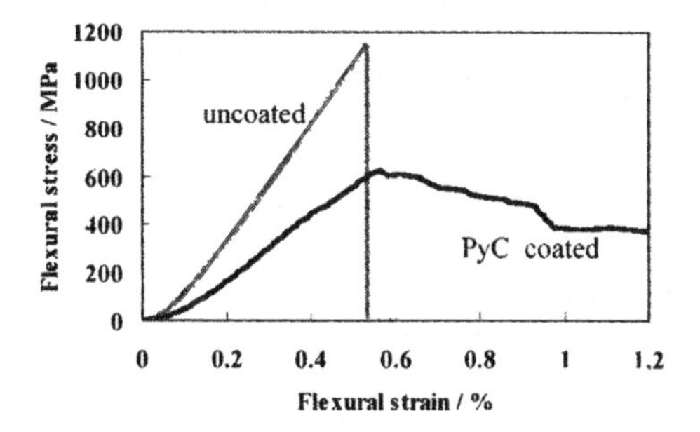

Fig. 8 Typical flexural stress-strain curves of SiC/SiC composites with uncoated and PyC coated fibers.

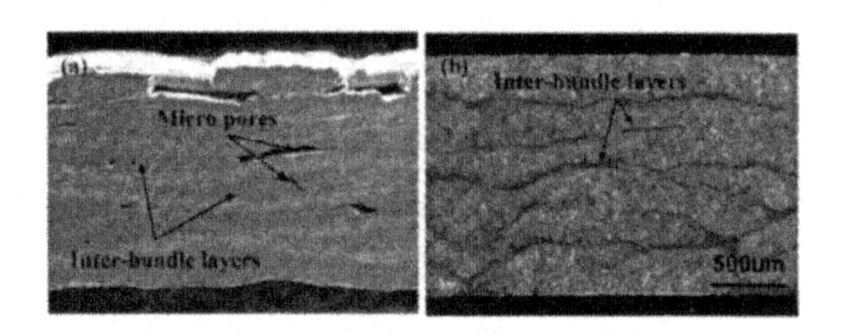

Fig. 8 Typical cross-section of SiC/SiC composites fabricated by: (a) CVI process (b) NITE process.

(~hundreds microns in size) at inter-fiber-tows and closed micro-pores (~ a few microns) at intra-fiber-tows are mainly observed in CVI and PIP composites, on the other hand distributed micro-pores (below 10μm) at inter-fibers are only observed in NITE composites.

SUMMARY

Matrix densification mechanism was proposed and discussed for the case of NITE process based on the following three parameters: (1) capillary adhesion force, (2) gas generation and (3) grain growth. Use of nano-sized powder is effective to increase capillary adhere force, resulting in denser matrix below 1900□ than submicron-powder. Additional SiO_2 enables lower temperature and viscosity of liquid phase, and thus increases capillary adhere force. Furthermore, Additional SiO_2 content plays significant roles to control precisely gas generation of volatile species. Control of SiC grain growth can be controlled by nano-sized powder characteristics and processing additives. Process control (sintering temperature/holding time) is another key.

Unidirectional SiC/SiC composites were fabricated by NITE process using SiC nano-sized powder with small amount of processing additives (S.C. + 3wt%SiO_2) for matrix formation. Induced PyC interphase conditions strongly affect the density, microstructural evolution, and therefore dominate mechanical properties and fracture behavior.

ACKNOWLEDGEMENTS

This work was supported by fundamental R&D on advanced material system for Very High Temperature Gas-cooled Fast Reactor Core Structures, the program funded by Ministry of Education, Culture, Sports, Science and Technology of Japan.

REFERENCES

[1] B. Riccardi, L. Giancarli, A. Hasegawa, Y. Katoh, A. Kohyama, R.H. Jones and L.L. Snead, "Issues and advances in SiC_f/SiC composites development for fusion reactors," *Jouranl of Nuclear Materials*, **329–333,** 56–65 (2004).

[2] R. Naslain, "Design, Preparation and Properties of Non-Oxide CMCs for Application in Engines and Nuclear Reactors: An Overview," *Composites Science and Technology*, **64,** 155-170(2004).

[3] A. Kohyama, Y. Katoh, T. Hinoki, W. Zhang and M. Kotani, "Progress in the Development of SiC/SiC Composite for Advanced Energy Systems: CREST-ACE Program," *Proc 8th European Conference Composite Materials*, **4,** 15-22(1998).

[4] R.H. Jones, L. Giancarli, A. Hasegawa, Y. Katoh, A. Kohyama, B. Riccardi, L.L. Snead and W.J. Weber, "Promise and challenges of SiC_f/SiC composites for fusion energy applications," *Journal of Nucleur Materials,* **307–311,** 1057 -1072(2002).

[5] T. Nozawa, T. Hinoki, L. L. Snead, Y. Katoh and A. Kohyama, "Neutron irradiation effects on high-crystallinity and near-stoichiometry SiC fibers and their composites," *Journal of Nuclear Materials,* **283-287,** 1157-1162(2002).

[6] T. Hinoki, L.L. Snead, Y. Katoh, A. Hasegawa, T. Nozawa and A. Kohyama, "The effect of high dose/high temperature irradiation on high purity fibers and their silicon carbide composites," *J. Nucl. Mater.* **283–287,** 1157-1162(2002).

[7] D. P. Stinto, A. J. Caputo and R. A. Lowden, "Synthesis of Fiber-Reinforced SiC Compsotes by Chemical Vapor Infiltration," *American Ceramic Society Bulletin*, **65**, *2*, 347-350(1986).

[8] R. Naslain, R. Pailler, X. Bourrat, S. Bertrand, F. Heurtevent, P. Dupel and F. Lamouroux, "Synthesis of highly tailored ceramic matrix composites by pressure-pulsed CVI," Solid State Ionics, **141-142,** 541-548(2001).

[9] M. Leparoux, L. Vandenbulcke, V. Serin, J. Sevely, S. Goujard and C. Robin-Brosse, "Oxidizing environment influence on the mechanical properties and microstructure of 2D-

SiC/BN/SiC composites processed by ICVI," *Journal of the European Ceramic Society,* **18**, 6, 715-723(1998).

[10] T. Taguchi, N. Igawa, R. Yamada and S. Jitsukawa, "Effect of thick SiC interphase layers on microstructure, mechanical and thermal properties of reaction-bonded SiC/SiC composites," *ournal of Physics and Chemistry of Solids*, **66**, 2-4, 576-580(2005).

[11] T. Hino, E. Hayashishita, Y. Yamauchi, M. Hashiba, Y. Hirohata and A. Kohyama, "Helium gas permeability of SiC/SiC composite used for in-vessel components of nuclear fusion reactor," *Fusion Engineering and Design*, 73, 1, 51-56(2005).

[12] M. Kotani, A. Kohyama, K. Okamura and K. Inoue, "Fabrication of High Performance SiC/SiC Composite by Polymer Impregnation and Poylysis Method," *Ceramic Engineering and Science Proceedings*, **20,** 309-316 (1999).

[13] Y.W. Kim, J.S. Song, S.W. Park and J.G. Lee, "Nicalon-fibre-reinforced silicon-carbide composites via polymer solution infiltration and chemical vapour infiltration," *Journal of Materials Science*, **28**, 3866–3868(1993).

[14] A. Kohyama, S. M. Dong and Y. Katoh, "Development of SiC/SiC Composites by Nano-Infiltration and Transient Eutectic (NITE) Process," *Ceramic Engineering and Science Proceedings*, 311-318(2000).

[15] S. M. Dong, Y. Katoh and A. Kohyama, "Processing Optimization and Mechanical Evaluation of Hot Pressed 2D Tyranno-SA/SiC Composites," *Journal of the European Ceramic Society*, **23,** 1223-1231 (2003).

[16] Y. Katoh, A. Kohyama, T. Nozawa and M. Sato, "SiC/SiC composites through transient eutectic-phase route for fusion applications," *Journal of Nuclear Materials,* **329-333,** 1, 587-591 (2004).

[17] N. P. Padture, "*In Situ*-Toughened Silicon Carbide," *Journal of the American Ceramics Society,* **77**, 2, 519-523(1994).

[18] K. Shimoda, J. S. Park, T. Hinoki and A. Kohyama, "Influence of Surface Structure of SiC Nano-Sized Powder Analyzed by X-ray Photoelectron Spectroscopy on Powder Basic physical properties," *Applied Surface Science(in-submitting).*

[19] K. Shimoda, M. Etoh, J. K. Lee, J. S. Park, T. Hinoki and A. Kohyama, "Influence of Surface Micro Chemistry of SiC Nano-Powder on The Sinterability of NITE-SiC," Proceedings of The 5th International Conference on High Temperature Ceramic Matrix Composites (HTCMC5), p. 101-106, The American Ceramics Society, Ohio (2004).

OPTIMIZATION OF SINTERING PARAMETERS FOR NITRIDE TRANSMUTATION FUELS

John T. Dunwoody, Christopher R. Stanek, Kenneth J. McClellan, Stewart L. Voit.
Los Alamos National Laboratory
P.O. Box 1663
Los Alamos, NM, 87544

Thomas Hartmann
University of Nevada-Las Vegas
Harry Reid Center for Environmental Studies
4505 Maryland Parkway, Box 454009
Las Vegas, NV, 89154-4009

Kirk Wheeler, Manuel Parra, Pedro D. Peralta
Arizona State University
Department of Mechanical and Aerospace Engineering
P.O. Box 876106
Tempe, AZ, 85287-6106

ABSTRACT

One mission of the Advanced Fuel Cycle Initiative (AFCI) is to develop transmutation fuels (*i.e.* fuels capable of having their transuranic elements transmuted to a shorter-lived isotopes) in order to close the nuclear fuel cycle, thereby reducing: the U.S. inventory of civilian plutonium, the waste stored in geologic repositories, and the cost of nuclear waste management. Actinide mononitrides are potential transmutation fuel materials due to favorable properties such as high melting point, excellent thermal conductivity, high fissile density, suitability towards reprocessing, and good radiation tolerance[1]. In order to avoid the difficulties of working with actinide bearing materials, we have performed a processing study on ZrN. ZrN is isostructural (*i.e.* NaCl structure) with actinide nitrides, and is therefore a good surrogate compound. Furthermore, ZrN is a diluent in so-called "non-fertile" fuels, *i.e.* containing no uranium. The effects of sintering were investigated by varying both the sintering atmosphere and temperature in order to determine the effect on important properties such as density and nitrogen stoichiometry. The density and stoichiometry values were then compared to hardness and microstructure on similar samples. The results of these studies are presented in the context of the various issues associated with the design and fabrication of actinide-bearing nitride fuel forms.

INTRODUCTION

The purpose of the research conducted by the Advanced Fuel Cycle Initiative (AFCI) is to produce a proliferation-resistant fuel form for Minor Actinide (MA) transmutation. If we can burn down plutonium and reduce the long term radiotoxicity of the remaining waste (through burn down of MAs) we will reduce the load on underground, protected repositories such as Yucca Mountain. It is estimated that by the time Yucca Mountain is opened and ready to receive shipments of nuclear waste, enough additional waste will have been generated to require another similar repository[2].

To this extent, AFCI is currently working to address these issues through fuels development research. This effort is comprised of two main elements, so-called Series I and Series II. Series I consists of developing proliferation resistant processes and fuels that enable the destruction of plutonium in light water reactors or high temperature gas-cooled reactors. Series II is focused on developing fuel cycle technologies that greatly reduce the long-term radiotoxicity and heat load of the high level waste to be sent to a geologic repository.

Recently, nuclear fuel development work at Los Alamos National Laboratory (LANL) has focused on low- and non-fertile mononitride fuel forms that fall under AFCI Series II. Low fertile (reduced plutonium breeding) fuel forms utilize a depleted uranium matrix, while the non-fertile (no plutonium breeding) fuel form utilizes an inert matrix. Zirconium nitride is considered one of the most promising inert matrices to be used in a nitride fuel cycle for fast reactors due to its low neutron capture cross-section, high thermal conductivity, and thermodynamic stability at high temperatures[3]. Research and development was performed on surrogate fuel forms (i.e. substituting cerium, depleted uranium, etc. for actinides) in support of actinide bearing fuel processing. The fuel fabrication process consists of converting precursor oxide fuel powders to nitride solid solutions, then cold pressing pellets and sintering to final desired parameters, such as density, open porosity, and grain size. In an effort to reduce waste, exposure and handling of radioactive material, actinide surrogates such as cerium and dysprosium are used. In addition, fundamental understanding of the sintering behavior of ZrN, a component of non-fertile fuel, would provide insight into the development of the ZrN bearing fuel. An added benefit of surrogate research is less costly materials and faster processing time (e.g. a "hot" process can take up to five times longer to complete than a "cold" process). The main goals of the surrogate nitride research are to provide fuel fabricators with a process that minimizes oxygen and carbon levels in the fuel as well as produce a final fuel form that will maintain density and mechanical integrity when pelletized and sintered.

Of particular interest is the loss of americium during processing via volatilization. It has been proposed that americium volatilization could be suppressed by sintering under a nitrogen atmosphere[4]. It was unknown, however, how nitrogen would affect densification, stoichiometry and resulting microstructure of the sintered pellet. Carbon and oxygen contaminants remaining as unreacted components of the carbothermic reduction/nitridization (CTR/N) process, continue to be a concern in both nuclear fuels and surrogate pellets. Hydrogen, added to the sintering gas of interest, was investigated for its potential to continue the CTR/N process and any potential effects on physical characteristics. The work in this paper investigates the effects of sintering atmosphere and temperature on density and nitrogen stoichiometry, and how the variations of these properties affect the mechanical integrity and microstructure of zirconium nitride pellets.

EXPERIMENTAL

Commercial ZrN (-325 mesh), zinc stearate (0.2 weight percent as fine powder), and polyethylene glycol (0.2 weight percent as coarse powder typically of 8,000 molecular weight) were milled in 5g batches in SPEX mill with one milling ball for 45 minutes. The 5g batches were combined into one large batch that was continuously recharged by hand mixing new sub-batches for 30 seconds. Charges of approximately 5g each were then cold pressed using a single action 13mm stainless steel punch and die set. The pellets were pressed to 250 MPa and held for

30 seconds, then slowly ejected. Geometric density was calculated on the green pellets, which were then placed in a tungsten boat (typically three per furnace run).

Two pellets were covered with loose ZrN powder, and the third was left uncovered to investigate surface effects from different atmospheres. Sintering was performed under gettered atmospheres of nitrogen, nitrogen - 6% hydrogen, argon, and argon - 6% hydrogen. Pellets were sintered under each gas at 1300°, 1400°, 1500°, and 1600° C. In addition, nitrogen and argon were used to sinter pellets at 1100°, 1200° and 1700° C. Pellets were sintered in an oxide tube furnace modified with a molybdenum tube surrounded by a sheath gas (to prevent oxidation of the Mo tube). The same temperature ramp profile was used for all sintering runs.

The sintered pellets were visually inspected for defects such as cracking, endcapping or hourglassing, as well as surface effects of the sintering atmosphere on the uncovered pellets. Density measurements (via geometric and immersion techniques) were then taken. Sections from uncovered pellets were obtained from each sintering run for lattice parameter and phase constitution analysis using X-ray powder diffraction, least square lattice parameter refinement, Rietveld structure refinement, microstructure analysis using SEM, and analysis of mechanical properties using Vickers indentation. A subset of the samples was also analyzed for nitrogen and oxygen content using a LECO TC600 residual gas analyzer.

RESULTS AND DISCUSSION

Sintering Atmosphere Effects on Densification and Lattice Parameter

The zirconium nitride pellets obtained were typically of good visual quality (*i.e.* no endcapping, cracking, laminations, swelling, or hourglassing). As seen in Figure 1, higher densities, as determined by immersion technique, were observed for pellets sintered under argon than those sintered under nitrogen, with the greatest difference of 17% of theoretical density at both 1400° and 1500° C. The peak density achieved for pellets sintered under argon was higher than that achieved for pellets sintered under nitrogen (92% vs. 85% of theoretical density), though we have not yet reached the maximum of the N_2 sintering curve. Pellets sintered under hydrogen containing atmospheres were lower in density than pellets sintered under the hydrogen-free analogs. Pellets sintered under either nitrogen or argon appear to follow the typical S-shaped sintering curve found in ceramics[5].

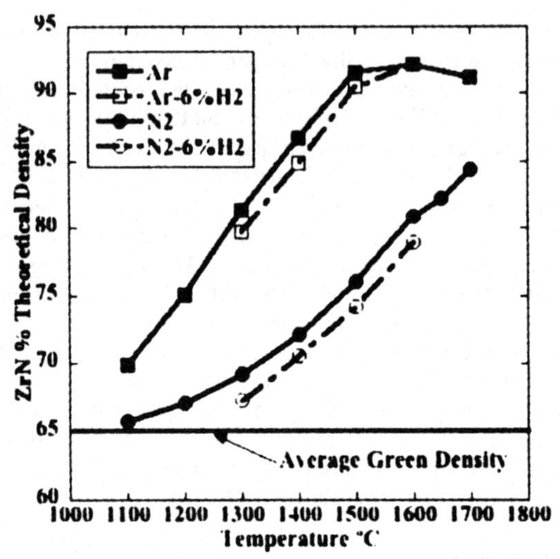

Figure 1. Densification trends by atmosphere and temperature.

A subset of samples sintered at 1300° and 1600° C under either nitrogen or argon, was analyzed for oxygen and nitrogen content. Nitrogen and oxygen weight percents were determined using a LECO TC600 oxygen/nitrogen analyzer. The results from these measurements revealed that pellets sintered under argon had a lower nitrogen weight percent than both as-received commercial ZrN and pellets sintered under nitrogen (12.3 wt. % vs. 12.6 wt. % and 14.7 wt. %, respectively). This finding was expected for the pellets sintered under argon, as it is conceivable that loosely bonded atomic nitrogen would be stripped away during sintering, the presence of atomic nitrogen being postulated by Straumanis et al.[6]. This is further supported by the noticeable mass loss, calculated as percent difference from green pellet mass, of pellets sintered under argon, as noted in Table I.

Table I Pellet properties across various sintering atmospheres and temperatures.

Analysis	1200°	1300°	1400°	1500°	1600°	1700°	Gas
Average % T.D.		67.33	70.59	74.20	78.94		N_2-6%H_2
Average % mass gain loss		1.34	1.27	1.21	1.24		
Average % closed porosity		2.90	2.96	2.91	3.13		
Average % open porosity		29.76	26.44	22.89	17.93		
Color of uncovered pellet		Silver-Gold	Silver-Gray	Silver-Gold	Dark Brown		
Average % T.D.	67.15	69.20	72.11	76.08	80.92	84.37	N_2
Average % mass gain loss	1.31	1.43	1.33	1.29	1.13	1.14	
Average % closed porosity	3.22	3.20	3.08	3.14	3.46	4.50	
Average % open porosity	29.63	27.60	24.80	20.85	15.62	11.30	
Color of uncovered pellet	Silver-Gold	Silver-Gold	Silver-Gold	Silver-Gold	Silver-Gray	Silver-Gray	

Average % T.D.		79.93	84.91	90.55	92.15		Ar-6%H$_2$
Average % mass gain/loss		-0.82	-1.14	-0.86	-0.77		
Average % closed porosity		4.16	5.20	7.56	5.93		
Average % open porosity		16.12	9.89	1.76	1.92		
Color of uncovered pellet		Silver-Gray	Brown-Gray	Brown-Gray	Dark Brown		
Average % T.D.	75.06	81.42	86.75	91.56	92.19	91.22	Ar
Average % mass gain loss	-0.67	-1.23	-0.85	-0.86	-1.02	-1.70	
Average % closed porosity	4.60	5.43	5.83	7.31	6.76	6.64	
Average % open porosity	20.34	13.16	7.42	1.14	1.05	2.14	
Color of uncovered pellet	Tan	Orange-Tan	Orange-Tan	Orange	Dark-Brown	Silver-Gold	

If nitrogen is lost during sintering in Ar, then it is expected that the nitrogen content of pellets sintered in nitrogen will be higher than those sintered under Ar, and possibly higher than the starting commercial powder. The nitrogen content of the commercial powder was measured to be hypo-stoichiometric. However, an interesting unexpected observation was the mass gain in pellets sintered under nitrogen (ZrN is reported not to accommodate hyper-stoichiometry with respect to nitrogen)[7]. Replacement of the oxygen with nitrogen would result in mass loss, with the removal of two oxygen atoms for every one nitrogen atom. The burnout of lube, binder, and absorbed punch lubricant would also result in a mass loss. Residual ZrO_2 can be converted to nitride under a nitrogen atmosphere in the presence of carbon[8]. It has been established that both ZrO_2 and carbon are present in the green pellet with additional carbon coming from the punch and die lube, pellet binder, and pellet lube, thereby potentially undergoing further carbothermic reduction/nitridization during the sintering process. This indicates that nitrogen is being added to the pellet through some undefined path (*i.e.* surface adsorption, incorporation into the lattice, etc.), zirconium is being removed from the pellet, or some combination of both. It is noted that oxygen content actually increased for all samples, regardless of temperature or atmosphere. It was also found through X-ray structure analysis that the pellets contained anywhere from 5-9% ZrO_2, with no apparent trend relating to sintering atmospheres or temperatures. Oxidation of the ZrN pellets through oxygen contamination in the sintering gas would, in effect, result in mass increase for all samples, a phenomenon not encountered in the pellets sintered under argon.

The variation of lattice parameter as a function of sintering condition was also investigated. Figure 2 shows the results of lattice parameter refinement of ZrN pellets sintered under various atmospheres and temperatures. It has been determined that the lattice parameter of ZrN is $a = 4.57560(5)$ Å[6]. (NOTE: Reported values of lattice parameters vary from 4.57560(5)[6] to 4.585(2)[9], dependent upon synthesis process (e.g. stoichiometry and impurity content). The value cited reflects similarities in material form and stoichiometry used and found in this study.)

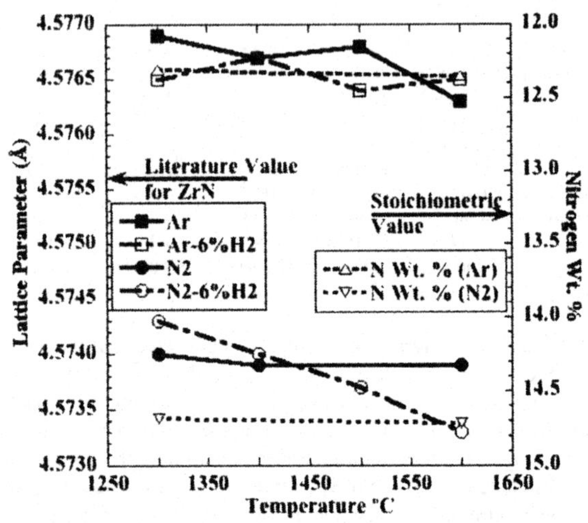

Figure 2. ZrN Lattice parameter development of ZrN samples in Å; the uncertainty of the lattice parameter refinement is approximately ± 0.0005 Å.

As can be seen in Figure 2, the determined lattice parameter for ZrN is higher than the refined lattice parameter determined using pellets sintered under nitrogen and lower than that found in pellets sintered under argon (a= 4.5739(3) Å and a= 4.5766(2) Å, respectively). If incorporation of excess nitrogen into the lattice were used to explain the increase in mass of the pellets, then it would be expected that the lattice parameter of the pellets sintered under nitrogen would increase. Conversely, nitrogen deficiency in the pellets sintered under argon could be expected to correspond to a reduced lattice parameter. Instead, it has been proposed that the higher nitrogen content is explained by zirconium vacancies, with all of the nitrogen lattice positions filled[6]. Removal of the relatively larger zirconium atom would indeed result in a shorter lattice parameter, thus supporting evidence of a Zr_xN structure, where $x<1$.

With all or most of the zirconium structure positions filled, substitution of O for N in the crystal structure would expand the ZrN unit cell, as seen in the pellets sintered under argon atmospheres. With no discernible changes in lattice parameter due to temperature, it is surmised that oxygen contamination in the sintering atmosphere is driving the change in lattice parameter for pellets sintered under argon. This phenomenon is compromised or eliminated in the presence of flowing nitrogen during sintering.

Sintering Atmosphere Effects on Microstructure

Covered pellets were consistently of the silver-gold coloration expected in sintered zirconium nitride pellets. Uncovered pellets were less consistent in color, overall, but trends in color were evident across a particular sintering atmosphere (see Table 1). Pellets sintered under nitrogen were silver-gold in coloration, with slight graying on the top and sides for the higher temperatures. This graying is possibly attributed to oxygen contamination in the sintering gas supply. Pellets sintered under nitrogen-6% hydrogen were also silver-gold, with the exception of the 1600°C run, which was dark brown in color. Pellets sintered under argon-6% hydrogen progressed from silver-gold through dark brown as the temperature was increased. Pellets sintered under argon showed the most diversity in color, progressing from a non-metallic tan through variations of orange up to dark brown. It appears that nitrogen as a sintering atmosphere has less of a surface effect on the sintered pellets than argon. Covering pellets with loose stock ZrN powder protects the pellets from these surface effects, which appear to be visual effects that do not correlate with trends in bulk properties.

Vickers indentation test results indicate that hardness of the ZrN pellets is affected by both sintering temperature and atmosphere. Hardness tests were performed on ZrN pellets sintered under nitrogen and argon, respectively, at 1300° and 1600° C. As can be seen in Figure 3, pellets sintered under argon are harder than pellets sintered under nitrogen. The temperature effect seen in the pellets show an increase in resistance to penetration with increased temperature, and thus, with increased density. The addition of hydrogen to a particular gas did not significantly affect the hardness of the sample for a given gas and temperature.

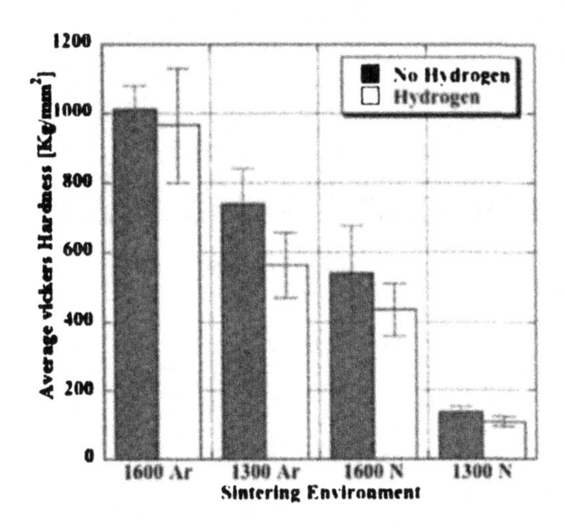

Figure 3. Hardness as determined by Vickers Indentation of ZrN sintered at 1300° and 1600°C: 500g load, 15 indents per sample

A comparison of SEM micrographs of the pellets sheds some light on this apparent disparity of hardness in the higher density ZrN pellet sintered under nitrogen. As seen in Figure 4, there is a drastic difference in microstructure between pellets sintered under argon versus those sintered under nitrogen at 1600° C. Pellets sintered under argon at this temperature, while being higher in density, also reveal a more equiaxed grain microstructure with uniform porosity and well defined grain boundaries. The pellets sintered under nitrogen at this temperature have regions of high density, with well defined (albeit smaller) equiaxed grains surrounded by a more porous region. Figure 5 indicates that this microstructure is similar to that of the pellet sintered under argon at 1300° C, at a similar density. From a fuel pellet design standpoint, the more equiaxed grain morphology and higher density would lead to better mechanical integrity, *i.e.*

resistance to cracking and thermal creep. However, if one considers that porosity, specifically open porosity, is intentionally designed into the fuel form in order to accommodate fission gas release, the microstructure seen in pellets sintered at 1600° C in argon would be inadequate, trapping the gas in the closed pores.

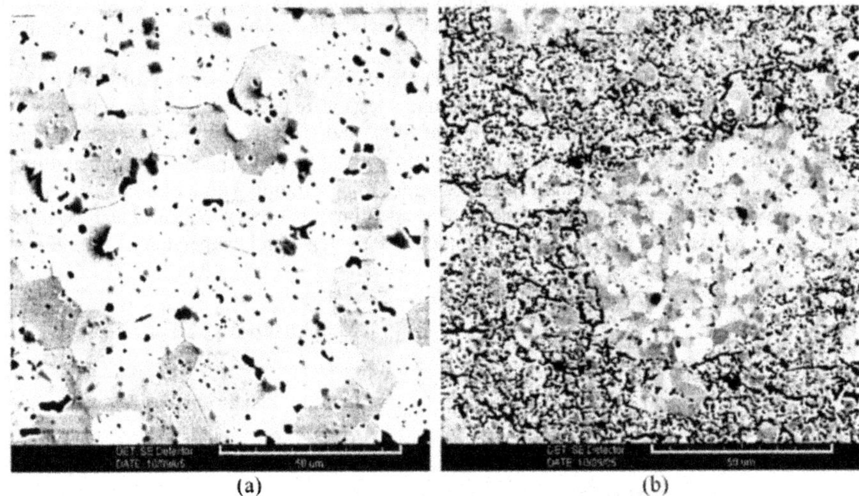

(a) (b)

Figure 4. Typical microstructure of ZrN sintered at 1600° C. (a) Argon atmosphere, (b) Nitrogen atmosphere. Scale bar is 50 μm in both images.

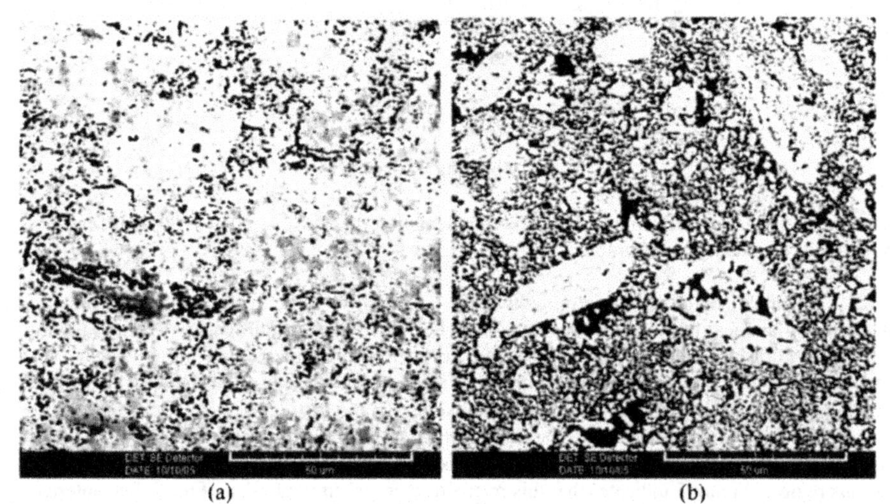

(a) (b)

Figure 5. Typical microstructure of ZrN sintered at 1300° C. (a) Argon atmosphere, (b) Nitrogen atmosphere. Scale bar is 50 μm in both images.

Another look at Table I indicates that while pellets sintered under argon at 1300° C may possess a density and microstructure similar to that of pellets sintered under nitrogen at 1600° C, there is a noticeable difference in the fact that the pellet sintered under nitrogen experienced a mass increase, whereas the pellet sintered under argon experienced a mass decrease. Another point of difference lies in the closed porosity of the two species. When comparing pellets of similar density, it is noted that pellets sintered under argon possess a higher percentage of closed pores. This closed porosity may be an indication as to the difference in mechanics of sintering at work under the two different atmospheres as differences in pore concentrations can be a result of differences in the rate of pore elimination near surfaces as an effect of temperature gradients experienced during the later stages of sintering[10]. The desired density (~ 85% of theoretical density) can be achieved using either nitrogen or argon as a sintering gas. The tradeoff comes in the other characteristics of the pellet. In general, argon would provide the desired hardness at a lower temperature, while nitrogen would suppress volatilization and provide a higher nitrogen content, the importance of which has not yet been ascertained.

CONCLUSION

Results of this study show that temperature and sintering atmosphere strongly affect the densification, microstructure, and nitrogen stoichiometry of zirconium nitride pellets. As a general trend, it is noted that densification of the pellets occurs at a lower temperature and achieves a higher value when sintered under argon, as opposed to nitrogen. The addition of hydrogen to the gas results in a lower densification of the pellets across the temperatures investigated while following the same densification curve as the parent gas. While the physical appearance of the pellets and oxygen content is similar across temperatures and gases, the pellets diverge in terms of mass gain or loss, nitrogen content, and lattice parameter when looking at nitrogen versus argon as a sintering gas. Pellets sintered under argon lose mass, have a hypo-stoichiometric nitrogen content and increase in lattice parameter when compared with pellets sintered under nitrogen. Lattice parameter, change in mass, and nitrogen content are strongly affected by atmosphere, while densification, hardness, and grain growth are affected by both atmosphere and temperature.

As these results relate to aiding in the development of actinide nitride nuclear fuel, information has been gathered on aspects of zirconium nitride behavior independent of actinide interaction. The complexities involved in forming a solid solution of the mononitrides involved in the fuel and the mechanisms of producing a viable fuel form under current physical constraints (finite time and temperature) are poorly understood. The effect of americium volatility suppression under a nitrogen atmosphere is decreased in utility due to the need for a much higher temperature to achieve the desired density. In addition, the equiaxed grain growth that occurs with increased densification diminishes the capability of fission gas release by reducing open porosity and increasing closed porosity. In addition to the results of this research, more characterization and studies of actinide interactions with the zirconium nitride matrix are needed to gain a better understanding of a final nitride fuel form.

ACKNOWLEDGEMENTS

This research was funded through the Advanced Fuel Cycle Initiative and NERI Contracts, under authorization of the Department of Energy.

REFERENCES

[1]A.A. Bauer, "Nitride Fuels: Properties and Potentials", *Reactor Technology*, **15**, No. 2 (1972).

[2]"Nuclear Waste Policy Act of 1982 (as amended)", U.S. Department of Energy, Office of Civilian Radioactive Waste Management, Washington, D.C., 20585, (March, 2004).

[3]M. Burghartz, G. Ledergerber, H. Hein, R.R. van der Laan, and R.J.M. Konings, "Some aspects of the use of ZrN as an inert matrix for actinide fuels", *J. Nucl. Matls.*, **288**, 233-236 (2001).

[4]M. Jolkkonen, M. Streit, and J. Wallenius, "Thermo-chemical Modeling of Uranium-free Nitride Fuels", *J. Nucl. Sci. Tech.*, **41**, No. 4, 457-65 (April, 2004).

[5]Principles of Ceramic Processing, J.S. Reed, 2nd Ed., Ch. 29, 583-624, John Wiley & Sons, Inc., New York (1995).

[6]M.E. Straumanis, C.A. Faunce, and W.J. James, "The Defect Structure Bonding of Zirconium Nitride Containing Excess Nitrogen", *Inorganic Chem.*, **5**, No. 11, (November, 1966).

[7]Phase Diagrams for Zirconium and Zirconia Systems, eds. H. M. Ondik and H. F. McMurdie, The American Ceramic Society, Westerville, OH (1998).

[8]M. Lerch, "Nitridation of Zirconia", *J. Am. Ceram. Soc.*, **79**, No. 10, 2641-2644, (1996).

[9]Pearson's Handbook Desk Edition, Crystallographic Data for Intermetallic Phases, P. Villars, Vol. 2 (CrSi-Zr), 2444-2445, ASM International, Materials Park, OH (1997).

[10]Introduction to Ceramics, W.D. Kingery, H.K. Bowen, and D.R. Uhlmann, 2nd Ed., Ch. 10, 448-515, John Wiley & Sons, Inc., New York (1976).

CERAMICS IN NON-THERMAL PLASMA DISCHARGES FOR HYDROGEN GENERATION

R. Vintila, G. Mendoza-Suarez, J. A. Kozinski, R. A. L. Drew
McGill University, 3610 University St. Montreal, Quebec, H3A2B2

ABSTRACT

The influence of ceramic specific capacitance (i.e. area, thickness and dielectric permittivity) on plasma intensity, which ultimately is correlated with the charge transferred in microdischarges, was studied in a dielectric barrier discharge reactor (DBD) at ambient conditions. The reactor operates at applied voltages ranging from 1-10kV and applied frequencies of 1-8kHz. The natural gas is injected in the plasma zone yielding hydrogen and solid carbon, with no CO or CO_2 release. This study showed that the increase in ceramic specific capacitance results in the increase in hydrogen yield and methane conversion rates due to the increase in the number of micro-discharges for higher dielectric constants. For dielectric constants ranging from 9 to 166 the hydrogen yield increased from 0.3% to 1.35%. In addition, the results suggest that CH_4 conversion rates of 2 and 6% were obtained for ceramics with dielectric constants of 9 and 166, respectively.

INTRODUCTION

Nowadays hydrogen and ceramics have being connected mostly through the substantial development of ceramic materials for hydrogen storage. In our work, ceramics and hydrogen have a different kind of association. The ceramics are, in fact, used for hydrogen generation via an environmental - friendly non-thermal dielectric barrier discharge process, dissociating natural gas.

Why hydrogen? The alarming concentration levels of CO and CO_2 resulting from burning fossil fuels as energy source has lead to the development of novel technologies towards alternative and clean sources of energy. Hydrogen, as a fuel, could provide the necessary answers for substantial green house gases (GHG) reduction since no CO or CO_2 are released from its utilization [1]. Paradoxically, the predominant, conventional method of hydrogen production, the steam reforming process, is accompanied by emissions of large quantities of CO_2 to the atmosphere [2]. Since *hydrogen is only as green as its source*, this clean fuel becomes not as clean after all due its fabrication process. This fact encourages development of nonconventional technologies such as plasma cracking of methane into hydrogen and carbon black [3-5]. Plasma processes, both in thermal and non-thermal regimes, have been given great attention in recent years. The most important direction of investigation is thermal decomposition. However, the main drawback of thermal plasma is the high-energy consumption resulting from the process high-temperature. Among different plasma regimes, the non-thermal plasma (NTP) is of great interest since chemical reactions can be induced employing much lower energies than in the thermal process. The electrical energy is employed to produce high-energy electrons rather than to increase the temperature of the surrounding gas [6]. Among the non-thermal processes, dielectric barrier discharge (DBD) plasma processes offer an attractive solution for hydrogen production. The advantage of an atmospheric pressure, the low temperature operation (350-400K) along with the presence of "hot electrons" make the DBD non-thermal plasma a high productivity process [7]. In DBD plasmas the dissociation of molecules is through the electron-impact, responsible for reaction initiation. Employing a dielectric in the current pathway between two metal electrodes creates "hot" electrons in the microdischarges, which initiate the chemical reactions. Without the DBD, only a few localized intense arcs would develop in the gas between the metal electrodes. Some operation parameters in DBD can easily be changed including electric factors such as applied voltage and frequency, as well as gas flow, discharge gap, electrode area and surface topography and dielectric ceramic permittivity or dielectric constant. The discharge characteristics such as plasma intensity or number of microdischarges and, consequently the yield products will change as the DBD parameters change.

The ceramic plays an important role in the proper functioning of the assembly [7], since the number of active species in the plasma, such as radicals and ions correlate to the permittivity or dielectric constant of the ceramic [8]. However, not many studies have been reported on the influence of the type of ceramic on the discharge characteristics of the reactor. The major problem that arises when using a commercial high-dielectric ceramic is that they tend to break in the plasma environment. The combination of high voltage, and the increasing reactor's temperature results in breakdown or failure of the ceramic. Consequently, the desired ceramic for this process should have a relatively high dielectric constant that increases with temperature and relatively good dielectric strength. The purpose of this study is to examine the influence of dielectric permittivity on plasma initiation voltage, plasma intensity (number of microdicharges), hydrogen yield, and methane conversion rates.

EXPERIMENTAL

Experimental Set-Up

Experiments were carried out in a DBD non-thermal plasma system (Fig.1) using a plasma dissociation chamber (PDC). A ceramic plate, which could vary in area, thickness and dielectric constant, is supported vertically in a Teflon cradle inside the PDC. On each side of the ceramic plate two metallic electrodes are placed in parallel position. The gap between the electrodes and the ceramic ranged between 0.25 -1.90 mm (10 - 75 thou). One of the metallic electrodes acts as a ground electrode while the other is connected to the plasma pulser through a step-up transformer. The applied voltage and frequency are monitored with a Tektronix TDS 320 oscilloscope, connected to a data acquisition system. The plasma discharge power is calculated by recording the reactor voltage and current, measured with a high voltage probe (Tektronix P6015A) and a current probe (Tektronix A6302), respectively. The plasma intensity, namely the number of microdischarges and their energy levels is monitored via a microdischarge analyzer. In each experiment helium was purged through the reactor (200 ml/min) in order to establish reference conditions. On one side of the reactor, a quartz window was placed for plasma observation. The reactor temperature is monitored by two K-type thermocouples placed near the ceramic plates and the exit.

A high voltage is applied between the electrodes producing a plasma discharge in the gap (Figure 2). When the plasma is generated, the natural gas is fed into the reactor, through plasma at 10-ml/min. The high-energy electrons produced in the microdischarges collide with the methane molecules leading to dissociation of the gas molecules into hydrogen and carbon. Formation of free radicals and other hydrocarbons in small amounts also occurs (e.g. C_2H_2). The carbon particles obtained in the dissociation process are retained on a paper filter.

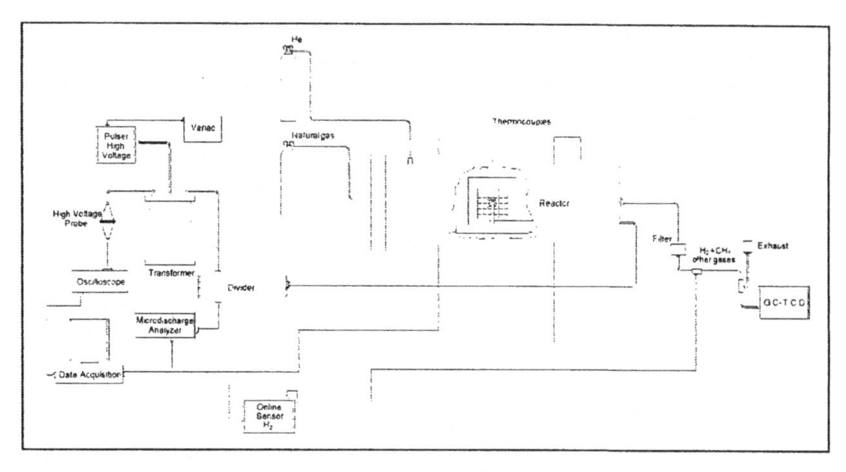

Figure 1 Experimental Set-Up

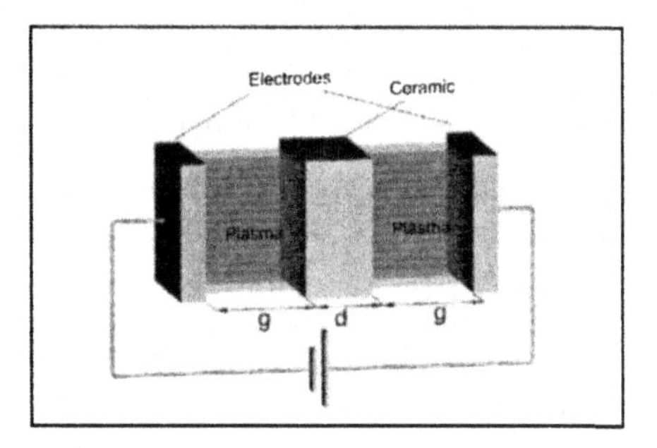

Figure 2 Geometrical arrangement of the DBD

The parameters varied in the present study are listed in Table 1. The degree of dissociation was measured at different discharge powers and dielectric media.

Table 1 Parameters varied in the process

Parameter		Range
Electrical	Voltage	1 - 10 kV
	Frequency	1 - 8 kHz
Geometrical	Discharge Gap	0.25 - 1.9 mm
	Electrode Area	0.30 - 14.30 cm^2
	Electrode Surface	Smooth - Coarse
Ceramic	Dielectric Thickness	3 - 8 mm
	Dielectric Constant	9 - 166

Gas and solid samples were collected at different stages of the conversion process. The composition of the evolving gases was determined using a Gas Chromatograph with a Thermal

Conductivity Detector (GC/TCD). The exhaust gases from the PDC reactor were continuously fed into the sampling loop. which was connected to a data acquisition system. The concentrations of H_2, CO, CO_2, and CH_4 in the outlet gases were determined on-line with a Nova 425 Analyzer.

Ceramic Compositions

Ceramic samples were prepared employing the conventional ceramic sintering technique. The powders were obtained courtesy of Ferro Corporation. The ceramic compositions and their properties are presented in Table 2.

Table 2 Ceramic compositions

Ceramic	Composition (range wt%)								Dielectric Constant
	BaO	TiO_2	ZrO_2	PbO	SiO_2*	CdO	CaO	Al_2O_3	
K-9	0	0	0	0	< 0.1	0	0	99.9	9
K-60	20-30	20-30	1-5	1-5	1-2	0	0	0	60
K-110	65	33	0	4	0	0	0	0	110
K-166	0	60	0	0	0	0	40	0	166

* amorphous, precipitated

Particle size evaluation was conducted in order to study the compactibility of the powder, followed by wet ball milling for 1 hour in polypropylene jars with zirconia media and 2wt% PVA binder solution. The weight ratio between milling media and powder was 2:3. The slurry was spray dried to obtain a uniform flowing powder and uniaxially pressed at 50 – 150 MPa into disks of 25-40 mm diameter and various thicknesses. The binder addition and the narrow particle size distribution provided the ceramic disks with enough green strength for handling. Sintering tests and density measurements were conducted to determine the optimum sintering profiles for the various compositions. The dielectric measurements were performed at frequencies varying from 1kHz to 1MHz on silver electroded disks using a high precision LCR meter (HP 4294A).

RESULTS AND DISCUSSION

In the DBD process a series of parameters could influence the optimization of the reactor. For this study only the influence of the dielectric ceramic - discharge gap assembly was considered.

Influence of Discharge Gap

The effect of discharge gap on plasma generation voltage (PGV), the voltage at which the first microdischarges appear in the gap, is shown in Figure 3 for the ceramic with dielectric constant $K = 166$.

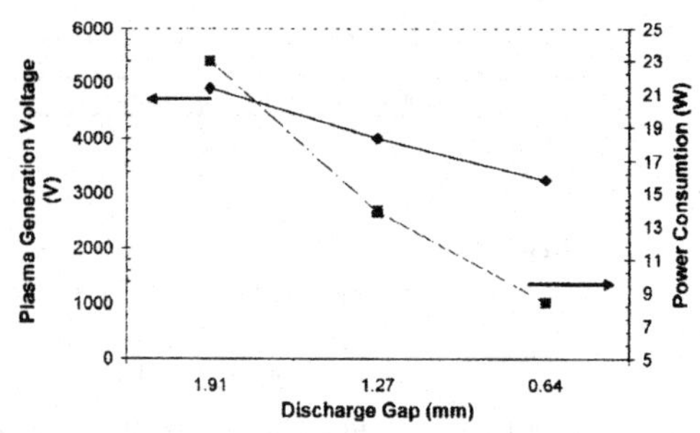

Figure 3 Discharge gap influence on PGV for K-166 ceramic at 2500 Hz and 1.22cm^2 electrodes

In the case of a 1.90 mm discharge gap, both the PGV as well as the power consumed for ignition presented the highest values. There is a decrease in PGV with the reduction of the gap, and so for the power consumed. A possible explanation of the onset plasma values decreasing with the discharge gap is related to the fact that the gas in the gap acts as a capacitor (equation 1).

$$C_g = \varepsilon_g S/d_g \qquad \text{(Equation 1)}$$

where C_g is gas capacitance, ε_g is the gas relative permittivity, S is the area of the electrodes and d_g is the gas gap. The smaller the thickness or the gap, the higher the capacitance and since the capacitance is proportional to the inverse of the voltage (equation 2), then plasma appears at lower voltage.

$$C = Q/V \qquad \text{(Equation 2)}$$

where C is the assembly capacitance, Q is the electric charge, and V the applied voltage.

Moreover the plasma generation versus gap was studied for different types of electrode surfaces, as illustrated in Figure 4.

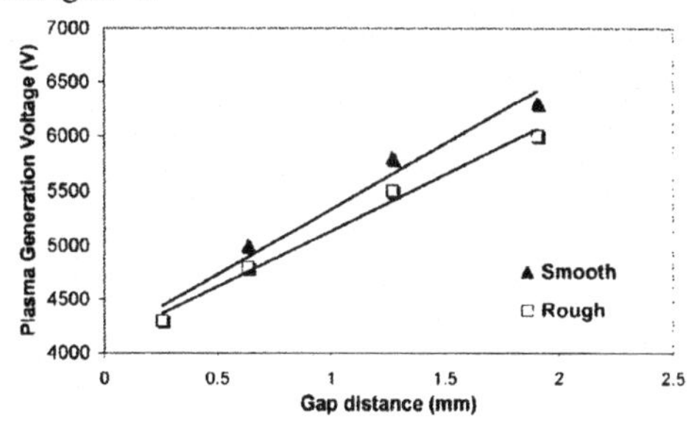

Figure 4 Influence of gap distance on PGV for K-9 ceramic for smooth and rough electrodes

An increase of PGV with gap increase was also noticed. In addition, the coarse surface (tiny pyramids) electrode generates plasma at slightly lower voltages than the smooth surface electrode. The concept is the same as a lightning rod and it is based on the principle that the electric field strength is concentrated around a pointed object, which ionizes the surrounding gas, enhancing its conductive ability. There is a conductive pathway established between the dielectric and the cathode that bridges the gap, which is where the microdischarge starts. Therefore, the microdischarges appear earlier, or at lower voltage values in a reactor with a coarse surface electrodes. The difference of PGV between the smooth and coarse electrode surface, increased as the gap widened. Consequently, for the experiments presented in this study, only the rough surface electrode was used.

Influence of Discharge Area (s)

The dependence between hydrogen concentration and the area of the discharge is shown in Figure 5. The increase in the discharge area results in an increase of the capacitance of the assembly since S increases, which in turn leads to the increase in the number of microdischarges resulting in higher hydrogen concentration.

At the same time, the increase in the discharge area could also be related to the increase in the residence time, which is 8 ms and 181 ms for RND2 (1.22 cm^2) and SQR1 (14.31 cm^2) electrodes, respectively. Longer residence times lead to longer residence times and higher levels of hydrogen concentration. Therefore the increase in hydrogen yield could be the combined effect of these two factors.

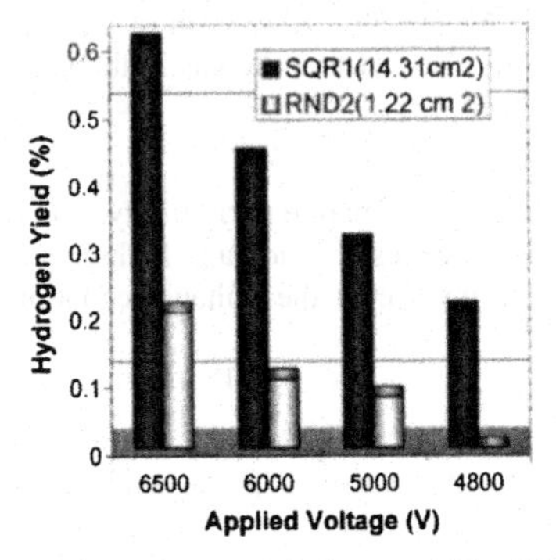

Figure 5 Influence of electrode area on hydrogen yield at 3500 Hz and various voltages

Influence of Dielectric Ceramic Thickness

The influence of dielectric ceramic thickness on hydrogen yield was measured for K-166 ceramic at different applied voltages (Figure 6).

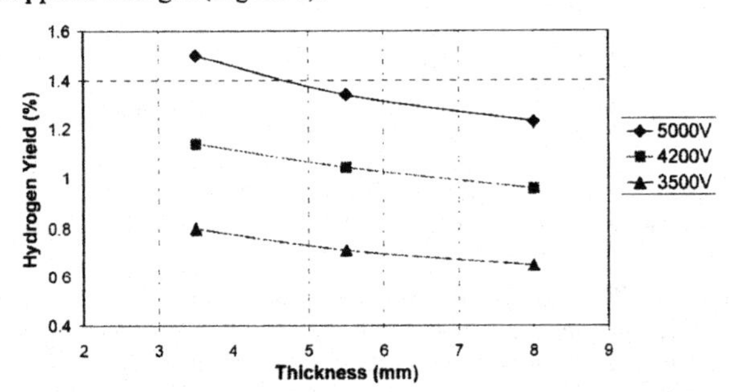

Figure 6 Influence of dielectric thickness for K-166 ceramic at different voltages

The increase in dielectric thickness resulted in a decrease of hydrogen yield. The total charge transferred into the microdischarges depends both on the gap width and the dielectric specific capacitance: area, dielectric constant, and dielectric thickness (equation 3).

$$C_d = \varepsilon_d S / d_d \qquad \text{(Equation 3)}$$

where ε_d is the ceramic relative permittivity, S is the area of the electrodes and d_d is the dielectric ceramic thickness. The increase in dielectric thickness results in a decrease of the capacitance of the active discharge reactor, which decreases the charge transferred to the microdischarges. Therefore, the hydrogen production is inversely proportional to the dielectric thickness.

Influence of Dielectric Constant

The investigation of the dielectric constant influence on the discharge characteristics of the reactor was the main objective of this study. In Figure 7 the effect of ε_d on PGV and on plasma intensity (number of microdischarges) is shown. The lowest PGV was found for K-166 among the four samples studied. The explanation resides in the fact that since the capacitance is inversely proportional to the applied voltage (equation 2): an increase in ceramic capacitance leads in a decrease of the onset voltage. However, the number of microdischarges is directly influenced by the increase of ε_d. This could be explained by the fact that the ceramic in DBD acts more like a capacitor and since the permittivity of the dielectric increases, the capacitance increases as well (equation 3).

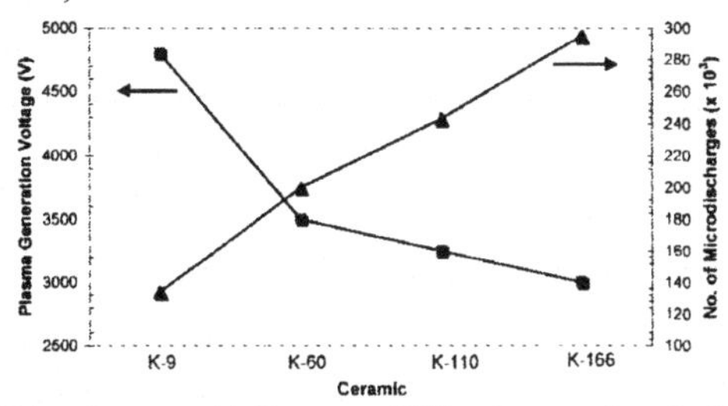

Figure 7 Dielectric constant influence on PGV and on number of microdischarges

At the same time, the potential energy stored in the capacitor increases and when the voltage changes polarity, the capacitor discharges with a higher intensity and this facilitates the formation of new microdischarges at the same spot before the previous microdischarge is fully dissipated, which leads to increased number of microdischarges.

The ceramic influence on current intensity was investigated as well (Figure 8). An increase in the discharge current with dielectric constant increase was found, which in turn reflects in the amount of charge transmitted to the microdischarges, and CH_4 conversion and hydrogen yield (Figure 9).

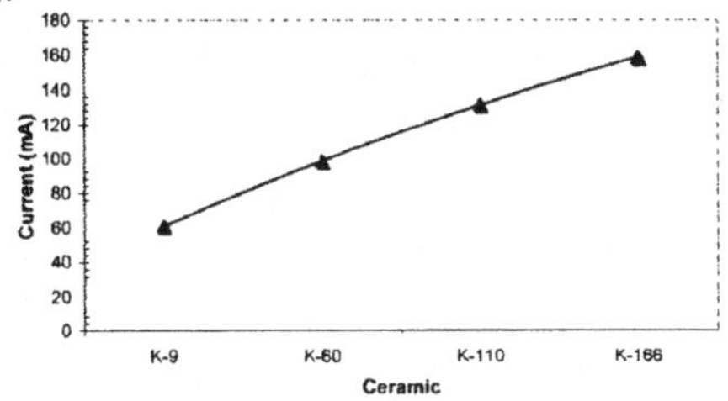

Figure 8 Dielectric constant influence on current intensity

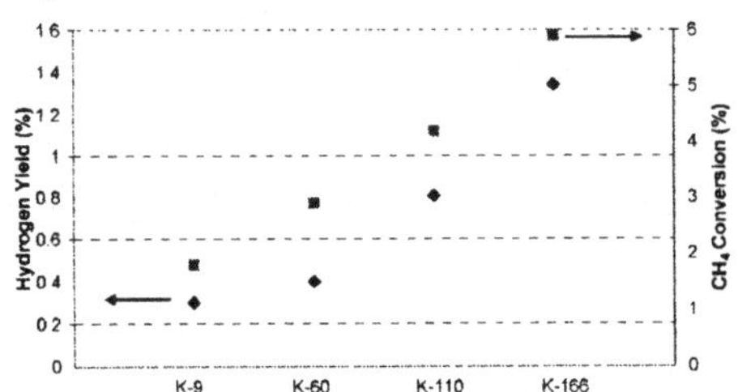

Figure 9 Influence of relative permittivity on hydrogen yield and methane conversion at 5000 V, 3500Hz, 0.63 mm gap

The hydrogen yield increased from 0.3% to 1.35% with the increase in the ceramic dielectric constant from 9 to 166, which is related to the larger production of electrons and ions. Therefore, the increase of the specific capacitance of the dielectric leads to a proportional increase in the charge transferred. Since the electric field strength is proportional to the amount of charge Q, which in turn is proportional to the capacitance and the applied voltage, the higher the capacitance the higher the stored charge, and hence, the electric field.

CONCLUSIONS

The specific capacitance influence of ceramics with permittivities varying between 9 and 166 on the number of microdischarges, onset voltage (PGV), current intensity, methane conversion rate and hydrogen yield, was experimentally investigated. The results show that the average number of microdischarges for a ceramic with K-166 ceramic is around 300×10^3. This value is almost 2.2 higher than that for the K-9 ceramic (135×10^3). The plasma initiation voltage also changed with the ceramic permittivity. It decreased from 4.8kV to 3kV by changing the dielectric permittivity from 9 to 166. In addition, current intensity increased with dielectric constant, which showed that the charge Q transferred in the microdischarges is influenced by the properties of the dielectric, which ultimately influences both the CH_4 conversion rates and hydrogen yield. Some control of the plasma characteristics was possible by varying applied voltage and frequency, adjusting the discharge gap or changing the electrode geometry and surface roughness and most importantly, by the properties of the ceramic dielectric constant.

REFERENCES

[1] Gaudernack, B. and S. Lynum, "Hydrogen from natural gas without release of CO_2 to the atmosphere", *International Journal Of Hydrogen Energy*, **23** (12) (1998) 1087-1093

[2] Fulcheri, L., Probst, N., Flamant, G., Fabry, F., Grivei, E., Bourrat, X., "Plasma processing: a step towards the production of new grades of carbon black", *Carbon*, **40** (2) (2002) 169-176

[3] Bromberg, L., Cohn, D. R., Rabinovich, A., Heywood, J., "Emissions reductions using hydrogen from plasmatron fuel converters", *International Journal of Hydrogen Energy*, **26** (10) (2001) 1115

[4] Fincke, J.R., Anderson, R. P., Hyde, T. A., Detering, B. A., "Plasma pyrolysis of methane to hydrogen and carbon black", *Industrial & Engineering Chemistry Research*, **41** (6) (2002) 1425-1435

[5] Fulcheri, L. and Y. Schwob, "From Methane To Hydrogen, Carbon-Black And Water", *International Journal Of Hydrogen Energy*, **20** (3) (1995) 197-202

[6] Kogelschatz, U., B. Eliasson, and W. Egli, "Dielectric-barrier discharges. Principle and applications", *Journal De Physique Iv*, **7** (C4) (1997) 47-66

[7] Eliasson, B. and U. Kogelschatz, "Nonequilibrium Volume Plasma Chemical-Processing", *Ieee Transactions On Plasma Science*, **19** (6) (1991) 1063-1077

[8] U. Kogelschatz, B. Eliasson, W. Egli "From ozone generators to flat television screens: history and future potential of dielectric-barrier discharges" *Pure Appl. Chem.* (1999) Vol. **71** No.10 1819-1828

PIEZOELECTRIC CERAMIC FIBER COMPOSITES FOR ENERGY HARVESTING TO POWER ELECTRONIC COMPONENTS

Richard Cass, Farhad Mohammadi, Stephen Leschin

Advanced Cerametrics, PO Box 128, Lambertville, NJ 08530

ABSTRACT

Recent advances in low power electronics have generated tremendous interest and provided new opportunities for the use of piezoelectric materials to power electronic devices without the need for batteries. Advanced Cerametrics (ACI) has refined its ceramic fiber spinning technology to produce piezoelectric fiber composites, which have generated sufficient power to run electronic devices from waste mechanical energy. Piezoelectric fiber composites were designed for maximum energy harvesting efficiency from ambient energy sources. These ceramic fibers are flexible, and in composite form, are very robust. Recent successes include powering a small (11 function) computer for over 100 seconds from10 seconds of a 25 Hz vibration (no battery). Energy harvesting devices, capable of producing and storing greater than 1 Joule of energy from 15 seconds of vibration will be presented along with ACI's smart sporting goods. Power generation from soft lead zirconate titanate (PZT-5A) piezoelectric fibers in the form of 1-3 and 1-1 composites, under application of an external force was investigated. 13, 45, 120 and 250 micron fibers were made by the Viscous Suspension Spinning Process and then formed into composites after sintering. Power generation experiments were performed and evaluated, with the best results being 400 volts pp and 120 mW of peak power.

BACKGROUND/INTRODUCTION: ENERGY HARVESTING

Energy harvesting (EH), sometimes referred to as energy scavenging, has gained tremendous attention as a means to lessen or eliminate the need for battery power. Novel approaches have been available to generate power by utilizing energy from human and environmental sources. Given battery limitations, wider adoption of EH is coming but requires a clever combination of design skills. A multi-discipline approach that leverages knowledge in several fields of engineering is required, including electrical, mechanical, material, and process. EH itself is not new. Consider hand-cranked radios, flashlights powered by shaking, windmill farms, and solar energy. What is new is the application of EH for ultra low power embedded electronics. The convergence of high charge piezoelectric ceramics and Advanced Cerametrics' ceramic fiber composite process technology has enabled the application of piezoelectric-powered systems for a wide array of electronic systems.

Piezoelectricity is the ability of certain crystalline materials to develop an electric charge proportional to a mechanical stress (direct effect), and a geometrical strain (deformation) proportional to an applied voltage (converse effect)[1]. The converse effect is used in piezoelectric actuators and makes possible quartz watches, micropositioners, ultrasonic cleaners, and an enormous number of other important products. A mechanical stress applied on a piezoelectric material creates an electric charge. Piezoceramics will give off an electric pulse even when the applied pressure is as small as sound pressure. This is called direct piezoelectric effect and is used in sensor applications such as microphones, undersea sound detecting devices, pressure transducers, etc. It is the direct effect, which is used in piezoelectric energy harvesting applications. Schematics of the direct and converse effects are shown in Figure 1. The generated charge is proportional to the external force pressure and disappears when the pressure is withdrawn. Some examples of the piezoelectric effects in a sampling of product developments are shown in Table I.

CONVERGENCE OF TECHNOLOGIES: ULTRA LOW POWER ELECTRONICS

The science of piezoelectric devices is fairly well understood in the engineering world, but their application remains a nascent field rich with possibilities. The emergence of piezoelectric ceramic fiber composite transducers that offer increasing deliverable power, combined with electronic components that measure performance in nanowatts, is opening a wide range of new product and services. The need for Extreme Life Span Power Supplies (ELSPS) for numerous electronics systems and devices is fueling extensive research, development, and growth. Piezoelectric ceramic fibers, given their unique properties, of flexibility, light weight and higher output per pound of material offer the greatest potential for enabling the wide-scale deployment of self-powered systems.

CONVERGENCE OF TECHNOLOGIES: PIEZOELECTRIC CERAMIC FIBERS FROM VSSP

Conventional piezoelectric ceramic materials are rigid, heavy, and produced in block form. ACI's low-cost piezo fiber forming technology, termed the Viscose Suspension Spinning Process (VSSP), can produce fibers that range in diameter from 10-250 μm (Figure 2).[2, 3] When formed into user defined (shaped) composites, the ceramic fibers possess all the desirable properties of ceramics (electrical, thermal, chemical) but eliminate the detrimental characteristics (brittleness, weight). The VSSP generates fibers with more efficient energy conversion than traditional bulk ceramics due to their, essentially, large length to area ratio ($V = g_{33}F\,L/A$, where V is piezoelectric generated output voltage, g_{33} is piezoelectric voltage coefficient, L and A are fiber length and cross-section area, respectively[4]). To put this into perspective, mechanical to electrical transduction efficiency can reach 70% compared with the 16-18% common to solar energy harvesting. And vibrations can be harvested 24 hours per day. The VSSP can produce fibers from almost any ceramic material.

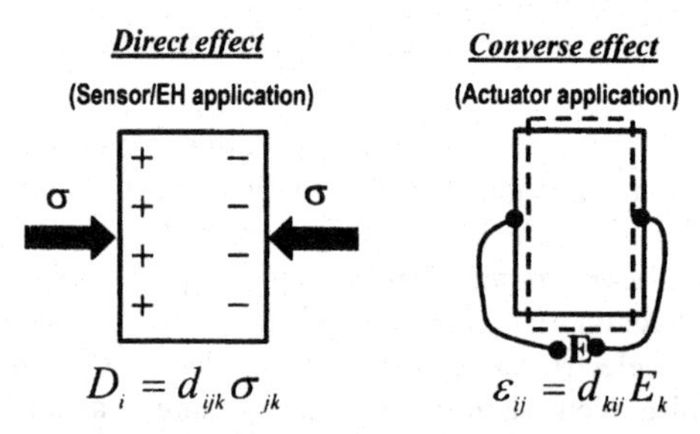

Figure 1. Schematics representing piezoelectric direct and converse effects.

Table I. Piezoelectric effects in various applications.

Application	Direct Piezo Effect	Inverse Piezo Effect
Energy Harvesting		
Battery Replacement	√	
Vibration Reduction		√
Reusable Energy Supply	√	
Condition Monitoring Sensors	√	√
Sporting Goods		
Smart Skis and Tennis	√	√
Racquets	√	√
Golf Equipment	√	√
Baseball bats, hockey sticks	√	√
Medical Devices		
Ultrasound Imaging	√	√
Self-Powered Pacemaker	√	
Defense		
Vibration Suppression		√
Active Structure Control	√	√
Sonar, Hydrophones	√	√
Ice Thickness Sensing	√	
Acoustical Devices		
Audio Reproduction	√	
Electronic Equipment	√	
Acoustic Suppression		√
Sensing		
Level & Weight Sensors	√	√
Non Destructive Testing	√	√
Smart Bearings	√	√

PIEZO POWER GENERATION

Advanced Fiber Composites (AFC) open the door for an array of energy harvesting applications. The fiber can recover (harvest) waste energy from mechanical forces such as motion, vibration, and compression (strain). With simple, low-cost analog circuits, the piezo power can be converted, stored, and regulated as a direct replacement for batteries. A typical single, AFC can easily generate voltages in the range of 40 $V_{p\text{-}p}$ from vibration. A typical single, AFCB (bimorph) can easily generate voltages in the range of 400 $V_{p\text{-}p}$ with some forms reaching outputs of 4000$V_{p\text{-}p.}$ Using a vibration frequency of *30 Hz*, ACI's piezo fibers have the proven the ability to produce 1 Joule of storable energy in a 13 second period enough to operate an LCD clock that consumes *0.11 mJ/s* for over 20 hours. Energy sufficient to power wireless systems for sensing and control of equipment, appliances, medical devices, buildings, and other infrastructure elements has been demonstrated.

Power output is scalable by combining two or more piezo elements in series or parallel, depending on the application. The composite fibers can be molded into unlimited user defined shapes and are both flexible and motion sensitive. The fibers are typically placed where there is a rich source of mechanical movement or waste energy. Other piezoelectric transducer form factor for energy harvesting is called piezoelectric multilayer composites (PMC). The properties of a typical PMC is shown in Table II. In PMC's the generated output voltage and power increases with increasing transducer thickness and decreasing fiber diameter, making small diameter piezoelectric fibers very attractive for energy harvesting (Figure 3).

Figure 2: The VSSP process: (a) typical spinning machine, (b) fiber formation at spinneret; (c) fiber take up, (d) spools of green spun fiber.

Table II. Properties of a typical PMC composite for energy harvesting.

Fiber type	PZT-EC65 (EDO)
Curie temperature (°C)	350
Fiber volume fraction (%)	30
Fiber diameter (μm)	13
Piezoelectric charge coefficient, d_{33} (pC/N)	375
Dielectric constant	700
Dielectric loss (%)	2
Coupling coefficient, k_t	0.65
Compliance, $S_{33}(D)$	2.93×10^{-12}
Compliance, $S_{33}(E)$	5.07×10^{-12}
Elastic constant $Y_{33}(D)$	3.42×10^{11}
Elastic constant $Y_{33}(E)$	1.97×10^{11}

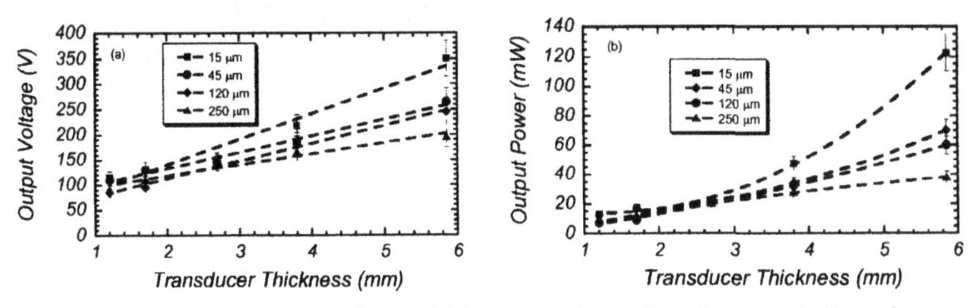

Figure 3. Effect of PMC transducer thickness on: (a) peak voltage, and (b) peak power. The transducers were made with various fiber diameters (15–250 μm).

EXAMPLE APPLICATION: WIRELESS SENSOR NETWORKS

Sensors that measure everything from process temperatures, to system pressures, to machine vibrations have been historically expensive to deploy in manufacturing and industrial environments. The sensors require expensive wiring and are expensive to service. With the emergence of the new Zigbee standard, based on IEEE 802.15.4, the availability of large, low-cost, low-power wireless sensor networks (WSNs) that are self-managed is becoming a reality. Sensors, signal conditioners, controllers, and RF transceivers continue to become smaller, lower power, and highly integrated. The combination of wireless networking, intelligent sensors, and distributed computing has created a new paradigm for monitoring the health of machines, buildings, and environments.

A low-cost, renewable energy source is critical to ubiquitous deployment of WSNs. After all, who wants to have to change thousands of batteries? New piezoelectric fiber-based energy harvesters, will in some cases, obviate the need for batteries in the WSNs. In other cases the harvesting technology can be used to recharge batteries to enhance service life. The power comes from the vibration of the system being monitored. Piezo fiber-based products will require no maintenance, significantly reduce the life cycle costs, and improve the overall quality of industrial and machine control systems.

Figure 4 shows an example of the PZT fiber acting as an energy harvester to convert waste mechanical energy into a self-sustaining power source for a Zigbee wireless sensor node. The piezoelectric fiber captures the energy generated by the structure's vibration, compression, or flexure. The resulting energy (current) is used to charge up a storage circuit that then provides the necessary power level for the sensor node electronics. In this example, energy is harvested by the vibration of PZT fiber composites. The energy is converted and stored in a low-leakage charge circuit until a certain threshold voltage is reached. Once the threshold is reached, the regulated power is allowed to flow for a sufficient period to power the Zigbee controller and RF transceiver.

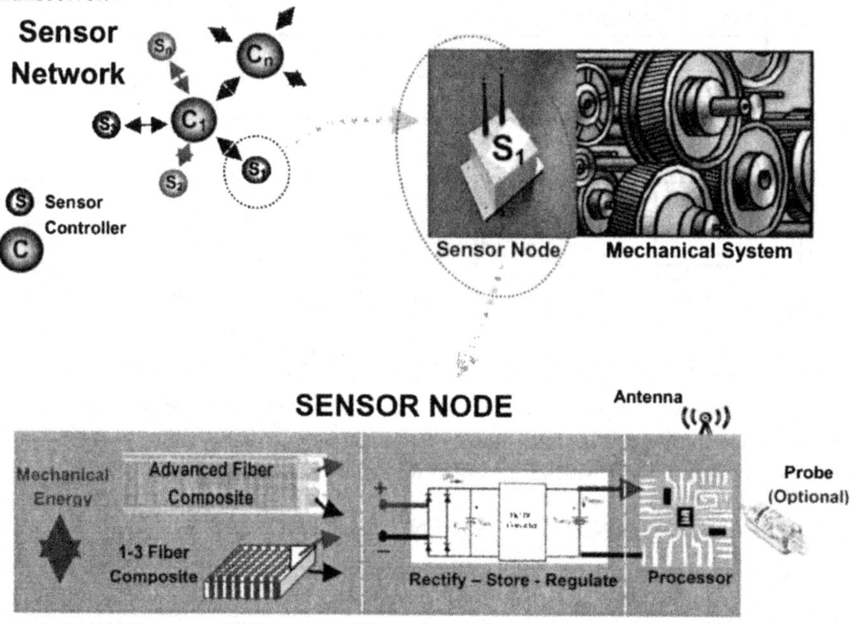

Figure 4 shows an example of the PZT fiber acting as an energy harvester to convert waste mechanical energy into a self-sustaining power source for a Zigbee wireless sensor node.

ADDITIONAL APPLICATIONS

Lighting

Advanced fiber composites can convert mechanical energy directly into light energy with no intervening electronics. By harvesting energy from ambient vibrations, advanced fiber composites can provide electroluminescent lighting on bridge decks, digital signage, buoys, and other low-power lighting loads.

Smart Structures

Advanced fiber composites also offer solutions for vibration damping and structural morphing. To enable self-adjusting systems, a smart structure containing advanced fiber composites senses a change in motion. The motion produces an electrical signal that can be sent to a control processor that measures the magnitude of the change in motion and returns an amplified signal that either stiffens or relaxes the advanced fiber actuators/sensors. Figure 5 demonstrates the application of advanced fiber composites transducers in powering various wireless transmitters.

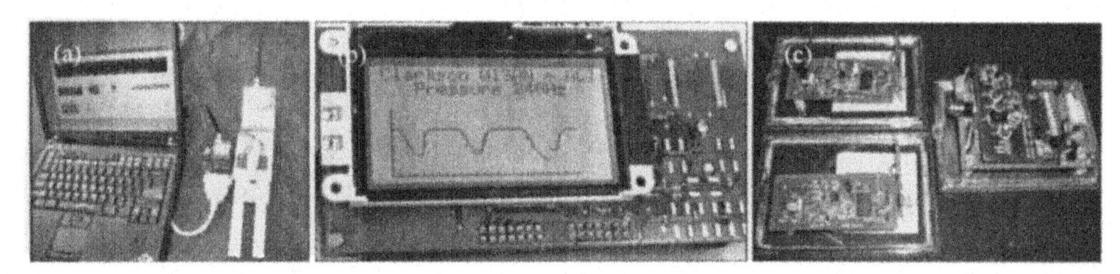

Figure 5: Examples of piezoelectric fiber powered wireless sensing systems: (a) transmitter/receiver, (b) pressure sensor receiver, (c) two encoded transmitters and a receiver.

SUMMARY

There are an emerging number of new and unique products coming to market that are limiting or obviating the need for batteries. Piezoelectric ceramic fiber technology provides a unique solution for EH, active structural control, and self-powered electronic systems. Advanced Fiber composites by ACI have 10x the power output of some other piezo forms, and have typical durability of 200 million cycles. By combining fiber composites with low-cost electronics and packaging, a new era of ultra low power products and applications is dawning. New market solutions are emerging that offer an ELSPS Factor for low power applications that is battery free.

ABOUT THE AUTHORS

Richard Cass, is the President of ACI.

Farhad Mohammadi is the Director of Research at ACI.

Steve Leschin is Managing Director of Business Development at ACI.

REFERENCES

[1] W.G. Cady, Piezoelectricity: An Introduction to the Theory and Applications of Electromechanical Phenomena in Crystals, New York: McGraw Hill, (1946).

[2] R.B. Cass, R.R. Loh, and T.C. Allen, "Method for Producing Refractory Filaments," *United States Patent*, No. 5,827,797, (1998).

[3] J.D. French and R.B. Cass, "Developing Innovative Ceramic Fibers," *American Ceramic Society Bulletin*, PP. 61-65, (1998).

[4] F. Mohammadi, A. Khan, R.B. Cass, "Power Generation from Piezoelectric Lead Zirconate Titanate Fiber Composites," *Materials Research Society Symposium*, V. 736, D5.5.1, (2003).

DESIGN FACTOR USING A SiC/SiC COMPOSITES FOR CORE COMPONENT OF GAS COOLED FAST REACTOR. I: HOOP STRESS

Jae-Kwang Lee[1], Masayuki Naganuma[1]

[1]Reactor Core and Fuel Design Group, O-arai Research and Development Center, Japan Atomic Energy Agency
Naritacho 4002, O-arai.machi, Ibaraki, Japan, 311-1393

ABSTRACT

As a part of the design study on Gas cooled Fast Reactor (GFR), core component designs of helium gas cooled fast reactor are being researched. The most promising structural materials of core components have been identified as ceramics, especially Silicon Carbide fiber reinforced Silicon Carbide matrix composites (SiC/SiC) have an encouraging characteristics. However, the existing design factors for Fast Breeder Reactor (FBR) were based on isotropic properties of metal. Therefore, the design factors of core components of GFR considered characteristics of ceramic composites was searched and investigated.

Hoop stress of assumed fuel pin model (cylindrical laminated composites tube using SiC/SiC) one of the design factors, it was estimated by thin or thick-walled pressure vessel theory and FEM analysis respectively. The stress of multi-directional laminates was changed at the interface of elements. If it assumes that each element are the ply of 2D weaving fiber, the changing of stress in the interface will act as laminar shear stress. Therefore, it was considered that to confirm inter-laminar shear strength of composites when we design fiber sequence of SiC/SiC composites is significant point. Through these work, reliability and safety of core component will be increased.

It was confirmed that applicability of SiC/SiC composites for fuel sub-assembly structural material. Feasible mechanical properties of unirradiatied SiC/SiC composites will have enough strength to the environment in the range of present study. These results are not only to design core components of GFR but also to indicate an improvement direction of SiC/SiC composites.

INTRODUCTION

The using of excellent thermal properties of ceramics as a structural material of nuclear field is being studied. The most promising structural materials being able to reach the ambitious goals of dose (> 100dpa) and temperature (> 1000°C) assigned to the Gas cooled Fast Reactor (GFR) core components have been identified as ceramics, especially carbides and nitrides [1]. However, ceramic material has inherent brittleness wherein their very poor toughness is the main drawback as a structural material and must be overcome. Therefore, Silicon carbide fiber reinforced Silicon carbide matrix composites (SiC/SiC) have been investigated to develop toughened ceramics in many fields [2,3]. Furthermore, Investigations of high temperature structural applications such as fuel sub-assembly of GFR are on going. Akira Kohyama reported about R & D of fabrication technology of fuel pins and core components [4]. Philippe Martin, et al. has been considered three different fuel concepts: a local confinement pellet fuel plate, a pin type fuel and a coated particles fuel [5]. Yasushi Okano et al. described GFR core design survey and selecting process on fuel configurations and, technical keys to achieve enhanced neutronics and thermal-hydraulics performance [6].

While SiC/SiC composites are relatively new materials with a limited database and design factors using SiC/SiC composites for GFR are not defined well. Design methodology requires sufficient understanding of their behavior in their application. In addition, the existing design factors for GFR were based on isotropic properties of metal. Therefore, this study was investigated and refined the design factors of core component which are considered characteristics of SiC/SiC composites.

STRENGTH OF FIBER REINFORCED COMPOSITES

Fiber sequence & laminating

Anisotropic characteristics of fiber reinforced composites distinguished from isotropic characteristics of metal caused by reinforcement mechanism. Composites arrayed fiber in uni-direction like Fig. 1 (c), it is just proper to uniaxial loading except a multiaxial loading. Real structures of industry generally use a quasi-isotropic material with laminating or weaving like a Fig. 1 (d), (e) and (f). 2D weaving fiber stacked composites should be most economical selection better than 3D weaving or lay-up of uniaxial fiber. Therefore, present study assumed that core component is made by 2D weaving fiber lamination.

Properties of continuous fiber reinforced composite should be controlled by the fiber strength, volume fraction, and matrix property. Therefore, the properties of composite will follow the rule of mixuteres which are described by Eq. (1) and it has commonly used in Fiber-Reinforced Plastic (FRP) study.

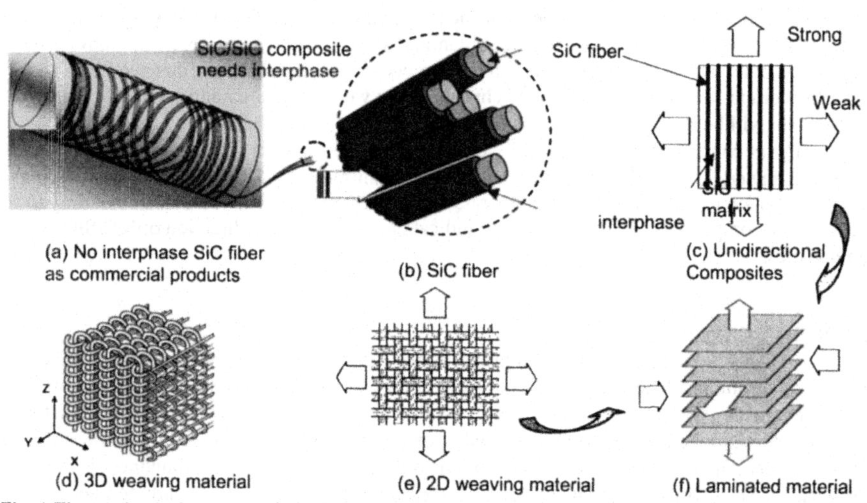

Fig. 1 The mechanical strength of SiC/SiC composites depends on fiber sequence and laminate method.

$$\sigma_c A_c = \sigma_f A_f + \sigma_m A_m \quad (1)$$

Where,

σ_c, σ_f, σ_m : Stress of composites, fiber, matrix, respectively

A_c, A_f, A_m: Area of composites, fiber matrix, respectively

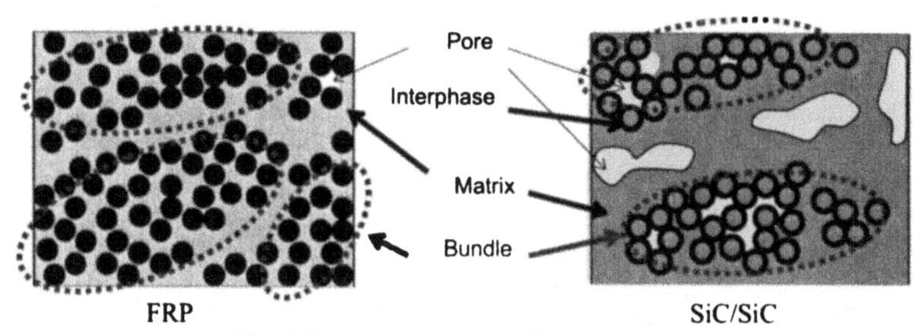

Fig. 2 Cross-section view of composites

Actually, it is possible to fine the efforts in literature about estimating for properties of a metal and ceramic based composite. [7]. However, It can not be applied to SiC/SiC composites directly. The properties of SiC/SiC composite do not satisfy simple relationship like an Eq. (1). It was demonstrated by comparison of composite properties. According to Eq. (1), the fracture strength of composite using a Nicalon® fiber should be 600 MPa for 40% volume fraction of 0/90 ° aligned fibers. But it was approximately a factor of three larger than the measured fracture strengths [8]. SiC/SiC composites are relatively new materials with a limited database, and fracture behavior to predict a strength of future composites was not clarified yet.

SiC/SiC composites (even though it shows a difference related to fabrication method) has relatively porous matrix better than FRP as shown in Fig 2. The rule of mixtures concept assumed that there is no pore. The effect of ununiformity of SiC/SiC composites matrix, far bigger than the matrix of FRP like an epoxy matrix, it can not be ignored. Furthermore, the characteristic of interface related to interphase and surface condition of fiber effects strongly on mechanical properties [9]. Therefore, it needs more experimental data and understanding for micro behavior of SiC/SiC composites to estimate the properties of future composite by rule of mixtures.

When design an engineering structure, the shape of cylindrical tube or vessel is a common structural element. The stresses can be normalized by the following stress quantities. It is assumed that there is perfect bonding between the layers of laminated tube. However, the stress distributions like a hoop and radial stress in laminated tube using a FRP are a function of the material properties, fiber orientations, stacking sequence, layer thickness, and tube aspect ratio [10].

Axial loading: average axial stress σ_o, where

$$\sigma_o = \frac{P_z}{\pi(R_o^2 - R_i^2)} \quad (2)$$

P_z : axial force
R_o: outer radius
R_i: inner radius
T_z : torque

Torsion: average laminate shear stress τ_o, where

$$\tau_o = \frac{3T_z}{2\pi(R_o^3 - R_i^3)} \quad (3)$$

These formulas also can not apply to SiC/SiC composites directly because of unignorable difference between FRP and SiC/SiC composites. Hence, it is difficult to express the hoop stress of cylindrical vessel using SiC/SiC composites as a normalized concept.

EXPERIMENT

Dimensions of assumed model

The dimensions of assumed cylindrical tube (like a fuel pin) using a SiC/SiC composites in Fig. 3 were determined from the design factors of existing FBR at the Table 1. It shows a value of D_o from 6 to 8.5 and D_o/w from 13.8 to 15.8. Dimensions of cylindrical tube were assumed that the outside diameter is 7mm and the ratio of D_o/w is 14, therefore w is 0.5mm.

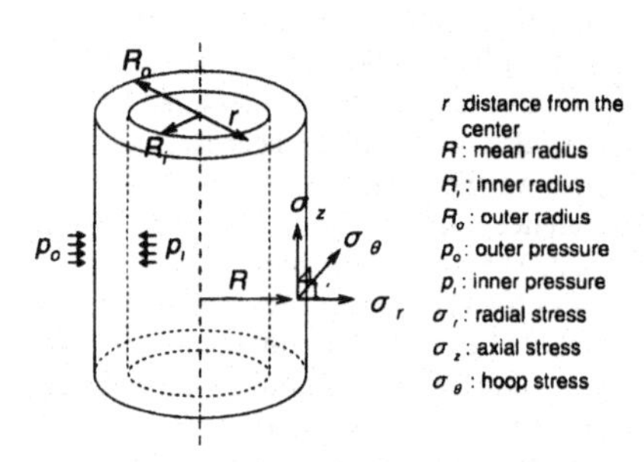

Fig. 3 Cylindrical tube

Table 1 Design factor of fuel pin for FBR [11]

	D_o^*	D_o/w	w^{**}
Monju (Japan)	6.5	13.8	0.47
Phenix (France)	6.5	14.6	0.45
PFR (U. K.)	5.84	15.4	0.38
SNR-300 (German)	6	15.8	0.38
Super-Phenix (France)	8.5	15.8	0.54

D_o^* : outside diameter of fuel pin, [mm]
w^{**} : wall thickness, [mm]

Material properties of SiC/SiC Composites

As an input data for stress estimation, feasible SiC/SiC composites properties are surveyed. For example, tensile strength of SiC/SiC composites is 200-500MPa by several fabrication processes and fiber orientations [12]. Assumed mechanical properties in present study are based on typical measured value of SiC/SiC composites in Table 2 [13]. Therefore, the hoop stress of fuel pin using SiC/SiC composites was estimated with assumed properties as shown in Table 3.

Table 2 SiC/SiC properties on present-day industrial composites

SiC/SiC properties and parameters[a]	Typical measured value
Density	≈ 2500 kg/m^3
Porosity	≈ 10%
Young's modulus	≈ 200 GPa
Poisson's ratio	0.18
Thermal expansion coefficient	4×10^{-6}/8 °C
Thermal conductivity in plane (1000 °C)	≈ 15 W/m K (BOL)
Thermal conductivity through thickness (1000 °C)	≈ 7.5 W/m K (BOL)
Electrical conductivity	≈ 500/Ωm (out of irradiation)
Tensile strength	300 MPa
Trans-laminar shear strength	200 MPa
Inter-laminar shear strength	44 MPa
Maximum allowable tensile Stress	Unknown[a]
Maximum allowable temperature (swelling basis)	≈ 1000 °C
Maximum allowable interface temperature with breeder	≈ 800 °C (static)
Minimum allowable temperature (thermal conductivity basis)	≈ 600 °C

[a] Assumed design criteria are slight different for each design. They are given in the appropriate chapters. No validated experimental data are yet available.

Table 3 Assumed mechanical properties of SiC/SiC

Parameter	Input data
Young' s modulus E1	200GPa
Young' s modulus E2	200GPa
Young' s modulus E3	100GPa
Poisson' s ratio ν12	0.18
Poisson' s ratio ν23	0.9
Poisson' s ratio ν31	0.9
Tensile strength (RT)	300MPa
Trans-laminar shear strength	200MPa
Inter-laminar shear strength	44MPa
Fiber sequence	0°, 45°, 0/45°
Wall thickness w, [mm]	0.5
Outer diameter D_o, [mm]	7.0
Inner diameter D_i, [mm]	6.5
Outer pressure (coolant pressure)	10MPa
Inner pressure (gas plenum)	0, 1MPa

STRESS EVALUATION

Thin & Thick-walled pressure vessel theory

GFR must maintain a certain amount of coolant pressure to carry heat effectively. That will bring a hoop and axial stress by pressure difference of the inner and outside of fuel pin. In the case of using composites, it is difficult to evaluate the hoop stress of fuel pin with theory for isotropic material. Strength and modulus of each direction has different value. However, if we evaluate the stress as a function of the material properties, it will be too complex and needs a lot of experimental data. Therefore, present study divided and calculated stress of the wall as some elements. Each element corresponds to ply of composites. The stress was calculated using following expression.

Thin-walled pressure vessel theory is commonly used in the evaluation of stress on the cladding tube by coolant and plenum gas [14].

- Thin-walled pressure vessel theory

$$\sigma_\theta = P\frac{R}{w} \qquad \sigma_z = P\frac{R}{2w} \qquad (4)$$

- Thick-walled pressure vessel theory

$$\sigma_\theta = \frac{p_i a^2 - p_o b^2}{b^2 - a^2} - \frac{a^2 b^2 (p_o - p_i)}{b^2 - a^2}\frac{1}{r^2} \qquad (5)$$

$$\sigma_r = \frac{p_i a^2 - p_o b^2}{b^2 - a^2} + \frac{a^2 b^2 (p_o - p_i)}{b^2 - a^2}\frac{1}{r^2} \qquad (6)$$

σ_θ: hoop stress
σ_z: axial stress
σ_r: radial stress
w: wall thickness
R: mean radius
R_i, a: inner radius
R_o, b: outer radius
P: pressure difference
p_o : outer pressure
p_i : inner pressure
r : distance from the center

To estimate a hoop stress by thin-walled pressure vessel theory is simple and comfortable obviously. However, some analysis codes with thin-walled theory assume that $\sigma_r = -P_i$; others ignore σ_r or use an approximation. Therefore, it is difficult to estimate the stress of laminated direction of composites. Thick-walled pressure vessel theory is commonly used in the evaluation of fuel deformation. It should be also useful to estimate a hoop stress of each position of fuel pin using a SiC/SiC composite.

Finite Element Modeling

Hoop stress was calculated using commercial FEM software "FINAS". Assumed fuel pin was simplified following figures. 0.5mm wall thickness was divided into 5 equal elements as shown in Fig. 4. Fiber sequence was simulated by coordinate rotation of element. In the case of quasi-isotropic laminates, it has 0/45/0/45/0° sequence.

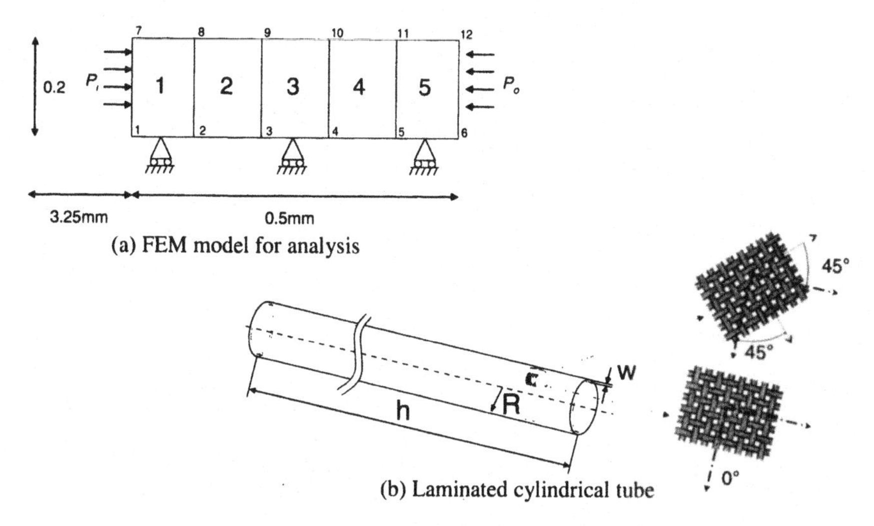

(a) FEM model for analysis

(b) Laminated cylindrical tube

Fig. 4 FEM model of SiC/SiC composites

RESULT & DISCUSSION

Evaluated hoop stress by each calculation method was compared in following Table 4 and Fig. 5. This result are the case of assumed model which have 10MPa (1.02kg/mm^2) coolant pressure and 1MPa (0.1kg/mm^2) plenum gas pressure. The stress of each position in 5mm wall thickness was compared. Minus value indicates compression stress. The stress of outer side ($(r-R_i)/w = 1$) was estimated higher than inner side ($(r-R_i)/w = 0$) stress by thin walled theory. However, the stress estimated by thick walled theory and FEM shows opposite result.

Table 4 Hoop stress by each evaluation method

$(r-R_i)/w$	Point	P_i [kg/mm^2]	P_o [kg/mm^2]	theory		FINAS		
				Thin [kg/mm²]	Thick [kg/mm²]	0° [kg/mm²]	45° [kg/mm²]	0/45° [kg/mm²]
0								
0.1	3.3	0	1.02	-6.732	-8.07318	-8.0655	-8.09279	-8.65697
0.3	3.4	0	1.02	-6.936	-7.8428	-7.82924	-7.84662	-6.98081
0.5	3.5	0	1.02	-7.14	-7.63188	-7.63421	-7.6369	-8.20099
0.7	3.6	0	1.02	-7.344	-7.43829	-7.45365	-7.4393	-6.62746
0.9	3.7	0	1.02	-7.548	-7.26019	-7.27051	-7.23518	-7.78607
1								

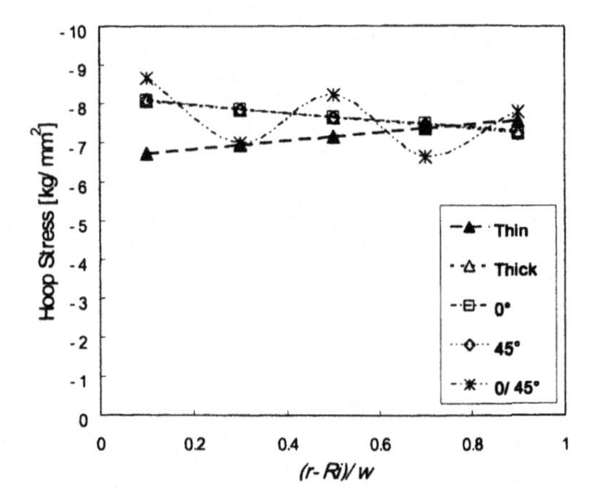

Fig. 5 Hoop stress by each evaluation method

In the case of uni-directional laminates, stress was just increased or decreased by the position from center point. However, the stress of multi-directional laminates was changed at the interface of elements. If it assumes that each element are the ply of 2D weaving fiber, the changing of stress value in the interface will act as laminar shear stress. Therefore, it was considered that to confirm inter-laminar shear strength of composites when we design fiber sequence of SiC/SiC composites is significant point. Through these work, reliability and safety of core component will be maintained.

SUMMARY

It was confirmed that applicability of SiC/SiC composites for fuel sub-assembly structural material. Feasible mechanical properties of unirradiatied SiC/SiC composites will have enough strength to the environment in the range of present study. Furthermore, another design factor for core component of GFR also will be studied and reported in future work.

- To estimate a hoop stress by thick-walled pressure vessel theory is more useful. It can be used when we estimate a hoop stress of core components like fuel pin using a SiC/SiC composite. Furthermore, FEM analysis also can be used when design fiber sequences of SiC/SiC composite.

- To realize a design concept of GFR, design work using CFCC (Ceramic Fiber reinforced Ceramic Composite) have to be refined more detail through reflect a progress of fabrication technology and evaluation method for SiC/SiC composites.

- To make a SiC/SiC composites real attractive material, fabrication technology and evaluation method have to reference and satisfy a requirements of design work like this study.

REFERENCES

[1]M. Dormeval, "Sintering and Characterization of Ceramics for GFR Applications", Proceedings of ICAPP05, Paper **5469** (2005).

[2]Y. Katoh, A. Kohyama, T. Nozawa, M. Sato, "SiC/SiC composites through transient eutectic-phase route for fusion applications", Journal of Nuclear Materials **329-333** pp. 587-591 (2004)

[3]S. Ueda, S. Nishio, Y. Seki, R. Kurihara, J. Adachi, S. Yamazaki, DREAM Design Team, "A fusion power reactor concept using SiC SiC composites", Journal of Nuclear Materials **258-263**, pp. 1589-1593 (1998)

[4]Akira Kohyama, "Advanced SiC/SiC Composite Materials for Fourth Generation Gas Cooled Fast Reactors", Key Engineering Materials Vol. **287**, pp. 16-21

[5]Philippe Martin, et al., "Gas Cooled Fast Reactor System: Major Objectives and Options for Reactor, Fuel and Fuel Cycle", Proceedings of GLOBAL 2005, Paper No. IL002

[6]Yasushi Okano et al., "Conceptual Design Study of Helium Cooled Fast Reactor in the "Feasibility Study" in Japan", Proceedings of GLOBAL 2005, Paper No. 412

[7]S.T. Mileiko, "Metal and Ceramic Based Composites", Elsevier, pp. 262-280 (1997)

[8]R.H. Jones etc al., "Radiation resistant ceramic matrix composites", Journal of Nuclear Materials **245** pp. 87-107 (1997)

[9]Francis Rebillat et al., "Properties of Multilayered Interphases in SiC/SiC Chemical-Vapor-Infiltrated Composites with "Weak" and "Strong" Interfaces", Journal of American Ceramic Society **81 [9]**, pp. 2315-2326 (1998)

[10]Carl T. Herakovich, "Mechanics of Fibrous Composites", John Wiley & Sons, Inc. pp. 362-401 (1998)

[11]IAEA LMFBR PLANT PARAMETERS (1991)

[12]A. Hasegawa et al., "Critical issues and current status of SiC/SiC composites for fusion", Journal of Nuclear Materials **283-287** pp. 128-137 (2000)

[13]L. Giancarli et al., "Progress in blanket designs using SiCf/SiC composites", Fusion Engineering and Design **61-62** pp. 307-318 (2002)

[14]Alan E. Walter, Albert B. Reynolds, "Fast Breeder Reactors," Pergamon Press Inc., New York, pp.278-285 (1981)

CHARACTERISATIONS OF Ti_3SiC_2 AS CANDIDATE FOR THE STRUCTURAL MATERIALS OF HIGH TEMPERATURE REACTORS

Fabienne Audubert
CEA Cadarache
DEN/DEC/SPUA
13108 Saint-Paul lez Durance, France

Guillaume Abrivard, Christophe Tallaron
CEA Le Ripault
DAM/DMAT/SR2C
37260 Monts, France

ABSTRACT

In the framework of an extensive R&D Program on High Temperature Reactors, the production and characterization of different ceramics liable to be used as structural materials are performed at the Commissariat à l'Energie Atomique (CEA). The Ti_3SiC_2 compound has received increasing attention due to its excellent properties in thermal conductivity and damage tolerance. The chemical stability and the evolution of the properties of the Ti_3SiC_2 compound under irradiation are a challenging issue. An irradiation program for the Ti_3SiC_2 compound is in progress. In this paper, we report the first characterisations of the Ti_3SiC_2 samples fabricated by reactively HIP sintering: thermochemical, thermal and mechanical properties. The objectives are (i) to determine exactly the properties of the compound which will be irradiated and (ii) to obtain data for fuel modeling codes. The thermal characterisations were highlighted the effect of the TiSi2 phase, as secondary phase into the Ti_3SiC_2 sample, on the thermal behavior of the compound. The inelastic behavior of the compound, even at room temperature, was confirmed. A test on Na-compatibility was carried out on the powder mixture with a DSC analysis and showed no reactivity between the Ti_3SiC_2 compound and the liquid Na.

INTRODUCTION

Thanks to the very unique helium properties as a coolant (almost transparent to neutron, no phase change, workable at very high temperature, ...), the High Temperature Reactors (HTR) are among the most attractive concepts for Generation IV nuclear systems. In such reactors, the temperatures achieved on internal structures are in the range of about 500°C to 1000°C in normal situation and can reach 1600°C in accidental situation. Such structures should be made of neutron transparent materials (in order to allow good performances of the core) allowing also for a refractory behavior. The materials are also submitted to different thermomechanical loadings (He pressure up to 70 bars, thermal gradients, displacement controled stresses according to the mechanical equilibrium of structures in contact ...). For these reasons, ceramics allowing for a good potential reliability, damage tolerance, thermal shock resistance, limited brittleness are primary candidates for structural materials. In this framework, it was decided to assess some MAX phases as possible candidates.

The Ti_3SiC_2 compound has received increasing attention due to its excellent properties in thermal conductivity and damage tolerance. Literature presents many works on the mechanical, thermal and oxidation behavior of Ti_3SiC_2 compound. On the other hand, no data exists on its behavior under irradiation. The knowledge on its chemical stability under irradiation (chemical transformation, decomposition, microstructural evolution) and on the evolution of its properties under irradiation (thermal conductivity, hardness and damage tolerance, irradiation swelling) is necessary to determine the potentiality of this compound for HTR application.

To evaluate the behavior under irradiation, two kinds of irradiation experiments can be performed: charged particles irradiation[1] and fast neutrons irradiation in nuclear reactor.

The Ti_3SiC_2 compound will be irradiated in fast neutrons reactor. In this paper, we report the first characterizations of the Ti_3SiC_2 samples fabricated by reactively HIP sintering: thermochemical, thermal and mechanical properties. The objective is (i) to determine exactly the properties of the compound which will be irradiated and (ii) to obtain data for fuel modeling codes.

EXPERIMENTAL DETAILS

Phase purity and microstructure

For this study, the Ti_3SiC_2 ceramics have been bought from 3-one-2® society. The samples were synthesized by pressureless sintering under an inert atmosphere, above 1500°C.

The sintered sample was crushed in liquid N_2 atmosphere and the powder was characterized using powder X-ray diffraction (D8 θ-2θ, Brüker). The composition was determined by chemical analysis performed at Laboratoire Central d'Analyse du CNRS (Vernaison) ; the carbon amount was obtained by infrared detection of CO_2 formed from solid burning, while the titanium and silicon amounts were determined by plasma emission spectroscopy. Optical microscopy (Provis AX70, Olympus) and scanning electron microscopy (SEM) (XL30, FEI) were used for microstructural evaluation. Qualitative analysis was performed by electron dispersive spectroscopy (EDS) (ISIS, Oxford Instruments). Sintered compact was etched using thermal treatment at 1550°C for 4 hours under flowing Ar. Sample density was obtained using the immersion technique.

Thermal properties

The thermal stability was evaluated on powders in a thermogravimetric associated to differential thermal analyzer (TGA/DTA) (TGA92, Setaram) and in a calorimeter (DSC) (Netzsch), and on small disks in a dilatometer (TMA92, Setaram) under a flow of Ar. The thermal expansion of bulk polycrystalline samples was measured under flowing Ar in the 25°C-1500°C temperature range in a dilatometer. The measurements were carried out on heating and cooling using a ramp of 3°C/min and a soaking time of 1 or 10 minutes. The thermal diffusivity was measured by a laser flash method, using a laser pulse (energy 40 J, duration 1 ms, wavelength 1 μm) directed on the front surface of the specimen located inside a high frequency induction furnace under argon atmosphere. The thermal diffusivity measurements were carried out over the range 430°C-1530°C. The thermal diffusivity results were converted to thermal conductivities using the heat capacity results and the measured density of the Ti_3SiC_2 sample used for the experiment. The heat capacity data were determined by the Netzsch society.

Mechanical properties at ambient and elevated temperatures

Four-point bending tests were performed on the polished Ti_3SiC_2 specimens with the size of 4 x 8 x 45 mm, at crosshead speed between 2 and 0.2 mm/min. The strain at ambient temperature was measured directly with an electrical-resistance strain gage mounted on the tensile surface of bending beams. The tests in temperature were performed on specimens with the size of 4 x 8 x 37 mm located inside a high frequency induction furnace under argon atmosphere. The temperature measurements were carried out using a pyrometer and two W-Re thermocouples located near the sample. The strain in temperature was measured using a displacement transducer.The statistical distribution of strength data, from 19 beams, using Weibull equation, was used to determine failure at ambient temperature. The sample surfaces and fracture surfaces were observed by scanning electron microscopy.

RESULTS AND DISCUSSION

Phase purity and microstructure

The mass composition of the solids (wt %) was 70.35 %Ti, 13.20 % Si and 12.17 % C (with respect to: 73.38 % Ti, 14.34 % Si and 12.28 % C theoretical). A contamination by metal compounds (about 0.1 wt % Co and Fe in particular) and by oxygen (about 1.2 %) was observed. The remaining is equivalent to the analytical error. These compounds could come from the thermal treatment. Figure 1 shows the XRD patterns of the Ti_3SiC_2 samples. The pattern shows peaks primarily belonging to the Ti_3SiC_2 phase, according to the XRD reported by Goto and Hirai[2], and also peaks belonging to TiC phase. Detailed analysis showed a small amount of $TiSi_2$. The lattice parameters (a=3.066 Å and c=17.630 Å) are comparable to those reported in the literature[3].

The SEM observation associated to EDX analysis (Figure 2) reveals that TiC phases were randomly localized in the whole sample (dark grey grains) and that a few grain boundaries were composed by $TiSi_2$ phases (light grey grains). The bright small particles were presumed to be Fe and/or Co contamination, according to chemical analysis. Figure 3 shows the microstructure of the Ti_3SiC_2 samples. As shown by the SEM micrograph, many small grains are surrounded by coarse plate-like grains, larger than 50 μm, which themselves are composed of layered structures or micro-lamellae. The average grain length was estimated to be 10 μm.

The bulk density was measured to be 4.51 g/cm^3, which is very near to the reported theroretical density of 4.53 g/cm^3 [4]. The high density was also confirmed by the SEM microstructural observations.

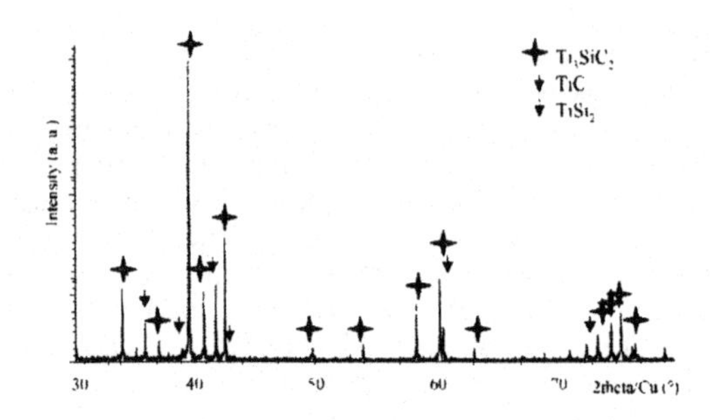

Figure 1: XRD diagram of Ti_3SiC_2

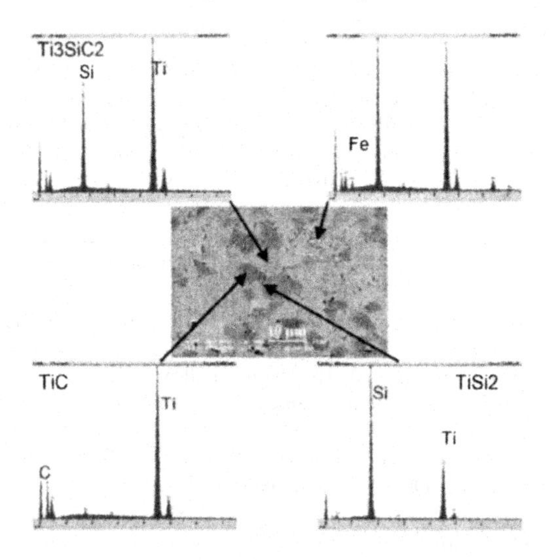

Figure 2: SEM micrograph and EDX analysis of the polished surface of Ti_3SiC_2

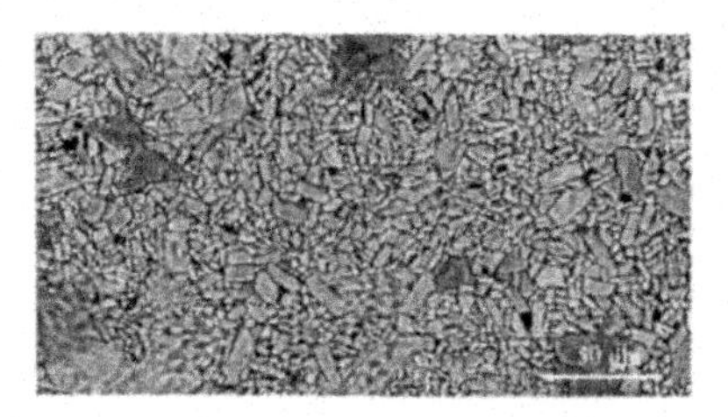

Figure 3: SEM micrograph of the polished and etched surface of Ti_3SiC_2

Thermal properties

Thermal treatment at 1500°C during 1 hour under a flow of Ar on a bulk sample leads to a higher purity of Ti_3SiC_2 sample. Indeed, the titanium carbide (TiC) and titanium disilicide ($TiSi_2$) peaks were no longer present in the diffraction pattern after annealing, which would indicate a progress in the synthesis reaction. However, the diffraction pattern reveals the presence of minor phases, like SiO_2 and Al_3Ti (alumina crucible) which were being elaborated by reaction with TiC and $TiSi_2$ phases (Figure 4). This hypothesis is being confirmed by occurrence of carbon peaks in the diffraction pattern. In agreement with literature[5,6], after a 4 hours treatment in presence of graphite, the resulting solid consists of a mixture of a larger amount of TiC and Ti_3SiC_2 but no $TiSi_2$ (Figure 5).

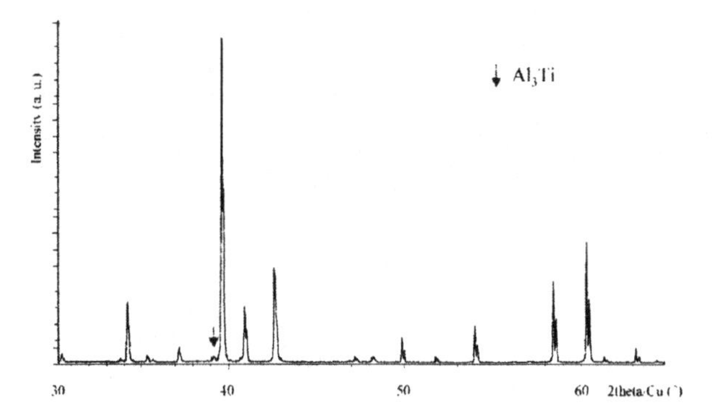

Figure 4: XRD pattern of a bulk sample of Ti_3SiC_2 calcined at 1500°C

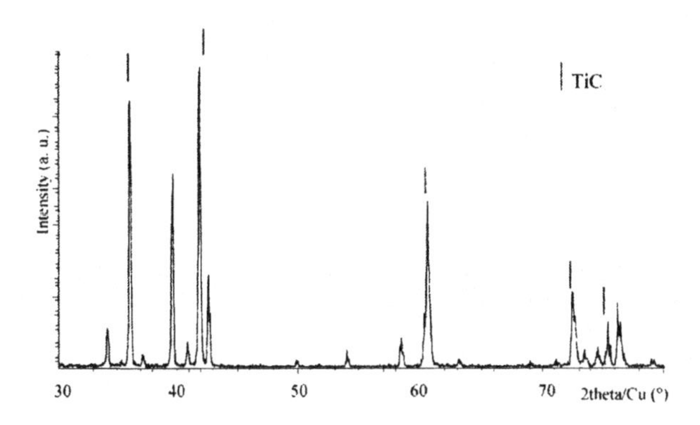

Figure 5: XRD pattern of a powder of Ti_3SiC_2 calcined at 1500°C in presence of graphite

Thermogravimetric analysis (TGA) shows an oxidation of the powder sample up to 1350°C (even in Ar atmosphere). Beyond 1350°C, no mass variations were detected up to maximum temperature evaluated, i.e. 1500°C. It is conceivable that a reaction occurs at 1350°C and that this one is opposed to oxidation reaction. Recorded under the same conditions as TGA, the corresponding differential scanning calorimetry (DSC) curves exhibits a small endothermic peak at 1330°C. A DSC analysis was carried out on the same sample a second time. The same result was obtained, indicating that the reaction is performed a second time. As shown by the phase diagram of the Ti-Si system, a liquid phase appears at 1330°C between $TiSi_2$ and Si. Be advised that the $TiSi_2$ phase appears together with TiC and Ti_3SiC_2, therefore it is possible to conclude the occurrence of a liquid phase $TiSi_2$-Si.

The dilatometric thermal expansion of a polycrystalline sample of Ti_3SiC_2, on both heating and cooling, is plotted in Figure 6. A least squares fit of the linear line (up to 1300°C) yields a slope of 10.3×10^{-6} °C^{-1} with an r^2 value of 0.999. This expansion value is a bit larger than previously reported values[7]. The curve presents a small shrinkage at 1350°C. The same curve was reported by Racault et al.[5] without explanation. The occurrence of a liquid phase, as mentioned previously, during heating, could lead to a liquid-phase sintering of the compound. It is worth noting that a permanent shrinkage (30 μm) was observed on the sample after thermal treatment. Several dilatometric treatments were carried out on the same sample and a decrease of the linear shrinkage values was observed.

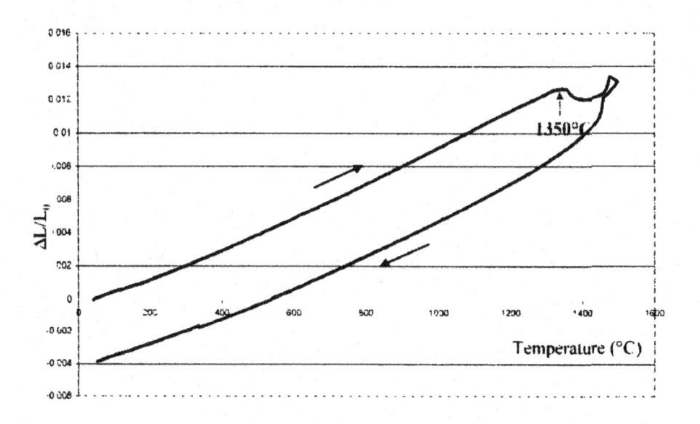

Figure 6: Thermal expansion of Ti_3SiC_2

The thermal diffusivity of Ti_3SiC_2 was measured as a function of temperature in the 430°C-1450°C range (Figure 7). Beyond this temperature, the sample was transparent to laser wavelenght (1 μm) and no measurement was carried out. This temperature corresponds to melting temperature of $TiSi_2$, and, for Racault and al.[5], to decomposition temperature of Ti_3SiC_2. The same values were obtained from another experiment with the same sample (covered by a graphite layer). The reproductibility of the values would indicate that the Ti_3SiC_2 phase is quite stable in these conditions. No explanation was found to understand these experimental results.

The thermal conductivity was evaluated from thermal diffusivity values, geometric density and experimental heat capacity. The temperature dependencies of the latter were determined up to 1200°C. The molar heat capacity increases with temperature up to 167 J/mol.K at 1200°C, a value sligthly higher than the value reported by Barsoum[7] (155 J/mol.K).

According to the literature, the thermal conductivity decreases with temperature. The values are slightly higher than the previously reported values[7]. A least square fit of the data yields the following relationship.

$$k\left[Wm^{-1}K^{-1}\right]=\frac{1}{4.26\cdot 10^{-6}T[K]+0.02355}$$

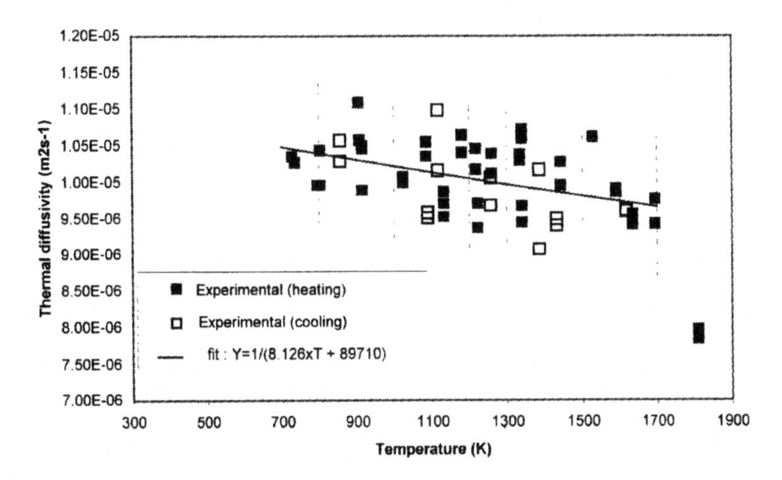

Figure 7: Temperature dependence of the thermal diffusivity of Ti_3SiC_2

Mechanical properties at ambient and elevated temperatures

The statistical distribution of strength data (Figure 8) yields to a Weibull modulus equal to 34, and a flexural strenght equal to 327 MPa (50 %).

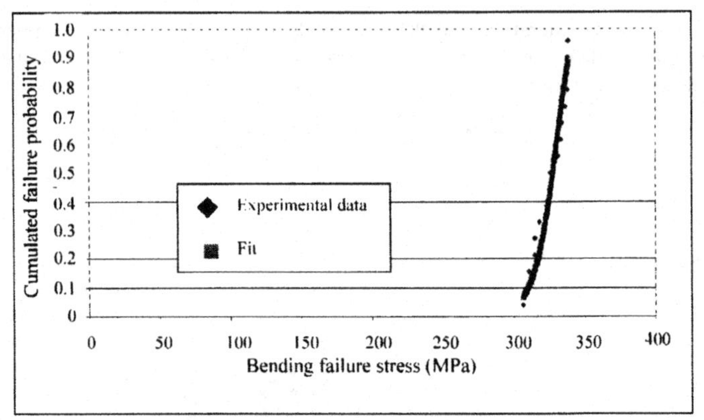

Figure 8: Weibull plot for the bending failure stress of Ti_3SiC_2

Figure 9a shows a representative load-strain curve measured at room temperature in a four-point bending specimen with a strain gage mounted on the tensile side (3 tests). A deviation from linearity was clearly observed before eventual fracture. Figure 9b presents a loading - unloading curve confirming the inelastic deformation of the sample. As suggested by Li et al.[8], the inelastic deformation behavior originates from the slip or shear deformation that is operative even at room temperature. The elastic modulus can be estimated simply using loading-unloading tests with a constant rate of the cross-head displacement and no-damage stress range. Despite inelastic deformation, the elastic modulus was evaluated from slope at the beginning. The modulus (average of 3 values) was determined as 283 GPa, slightly lower than the value reported in the literature (315 GPa)[9].

High-temperature flexural strenghts were determined at 800°C, 1000°C and 1200°C. Figure 10a shows the four-point bending load-strain curves measured at elevated temperatures. The curves obtained at temperatures below 1000°C are similar to that at ambient temperature and the specimens showed brittle fracture at the maximum load. At 1200°C, the specimen was easily deformed under a low flow stress. Figure 10b shows the variation in bending stress as a function of temperature. The average bending strength at ambient temperature was about 340 MPa and tended to decrease with increasing temperature. The values are slightly higher than ones reported by Li et al.[8] (reported on the Figure 10b).

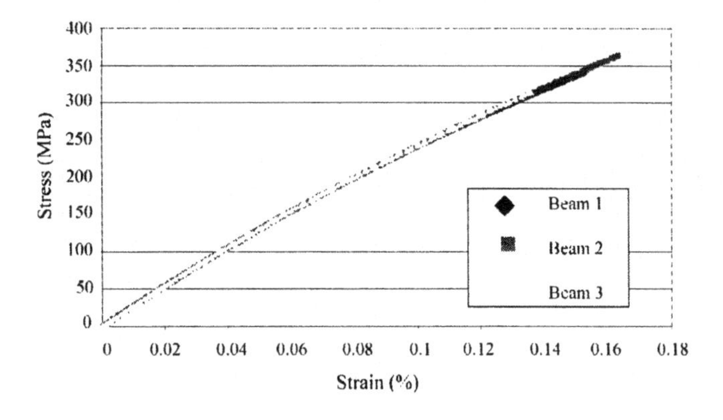

(a)

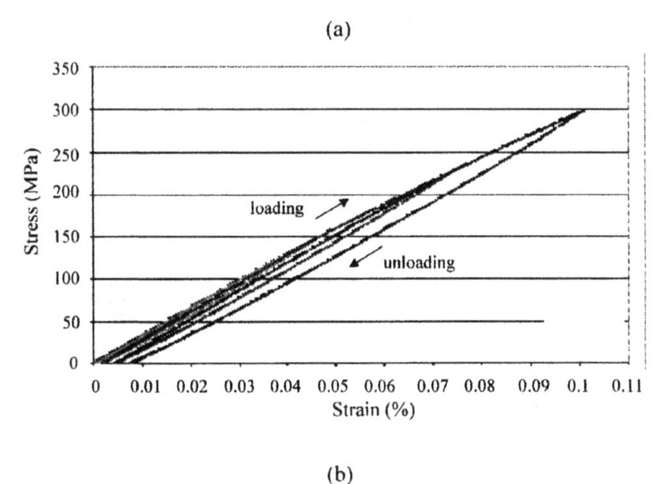

(b)

Figure 9: Ambient-temperature load-strain curves (a) and loadind-unloading curves (b) of Ti_3SiC_2

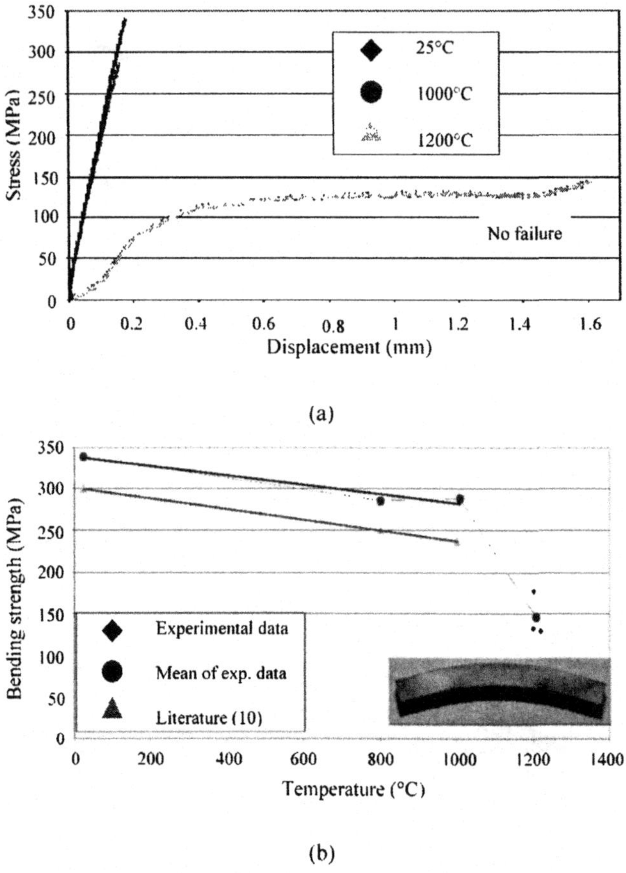

Figure 10: Load-strain curves obtained at elevated temperatures

A four-point bending test was carried out at 1200°C and was continued during temperature decrease. A brittle fracture was observed on the specimen which has been deformed at 1200°C, and the flexural strenght was determined as about 300 MPa, a value similar than one obtained at lower temperature than 1200°C. This result indicates that the damage formed by sample deformation does not modify the fracture mechanism and the bending strength. The comparison of several fracture morphology of Ti_3SiC_2 specimens at different temperatures reveals that no obvious difference was found in the fracture mechanism, i.e. brittle fracture mode. At 1200°C, the surface observations show that there are many micro-cracks (Figure 11). The fractographies of beams at different temperatures also show these microcracks inside the specimen. These observations are in good agreement with the micrographies reported by Zhang et al.[10].

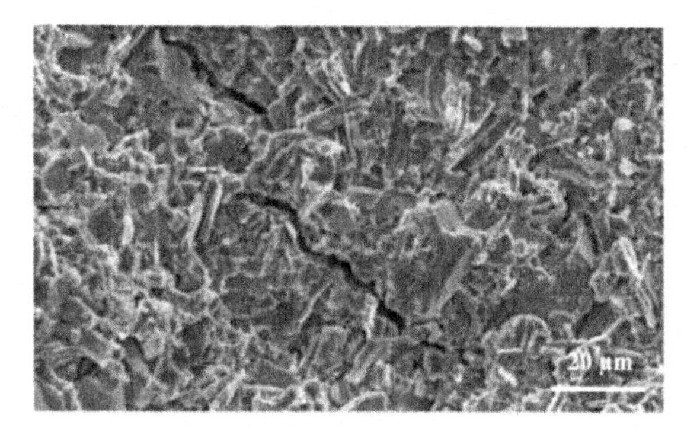

Figure 11: SEM micrograph of the surface of Ti_3SiC_2 at 1200°C

Irradiation behavior

Irradiation behavior of the Ti_3SiC_2 ceramic will be evaluated from fast neutrons irradiation tests in Phénix reactor (France) and in Osiris reactor (France). The irradiation experiment in Phénix reactor is foreseen in 2007, for a residence time of about 400 EFPD, i.e. 32 dpa. The temperature on Ti_3SiC_2 samples is expected at 500°C. The irradiation experiment in Osiris reactor was in progress since December 2005, for two years, with 1.4 dpa. This test was carried out at room temperature. The irradiated specimens are cylindrical; the microstructural features of each sample were determined and will be compared after irradiation tests. The macrostructural evolutions (swelling) will be determined.

To carry out irradiation test in Phénix reactor, it was necessary to evaluate Na-compatibility (possible interaction) with the Ti_3SiC_2 samples. A DSC analysis was carried out up to 1100°C for 5 minutes. The mixture of the Ti_3SiC_2 powder and liquid Na was introduced into a sealed crucible of inconel. No reaction was observed between the sample and liquid-Na. On the other hand, the curve shows that a reaction between Ti_3SiC_2 and Ni is possible. A DSC analysis on a powder mixture of Ti_3SiC_2 and Ni and an X-ray diffraction analysis on the mixture calcined at 1500°C were realized. These analyses confirm a reaction between the.compounds and show the decomposition of the Ti_3SiC_2 sample. Indeed, the diffraction patttern exhibits only the peaks of TiC phase and of Ni phase, no peak of Ti_3SiC_2 phase occurs. These experiments revealed that in presence of Ni, decomposition temperature of the Ti_3SiC_2 sample decreases.

CONCLUSIONS

The Ti_3SiC_2 compound was considered as a potential structural material for internal structures of High Temperature Reactors. Thermochemical characterisations were carried out on reactively HIPed Ti_3SiC_2 samples.

It was noted that the compound contains a small amount of TiC and $TiSi_2$ as minor phases. The occurrence of $TiSi_2$ plays an important role on the thermal stability of the Ti_3SiC_2 compound, in forming a liquid phase at 1330°C. Indeed, a liquid-phase sintering was observed during thermal expansion measurement combined with an endothermic peak observed by DSC analysis.

The four-point bending tests at different temperatures confirmed the inelastic deformation of the Ti_3SiC_2 compound. The Weibull modulus of Ti_3SiC_2 was determined at room temperature: m=34.

In order to allow for a fast neutrons irradiation test in the Phénix reactor, the no-reactivity between Ti_3SiC_2 and Na was confirmed.

ACKNOWLEDGMENTS

The authors wish to aknowledge Mathias Soulon for his valuable technical collaboration, Jean-Christophe Richaud, Virginie Basini and Jean-Pierre Ottaviani for thermal diffusivity and DSC experiments, and NETZSCH society for Cp measurements. The authors are grateful to Dr T. El-Raghy for fruitful discussion.

REFERENCES

[1]F. Audubert, G. Carlot, L. David, S. Gomez and J. Lechelle, "Inert materials for the GFR fuel: Characterizations, Chemical Interactions and Irradiation Damage", *Proceedings GLOBAL 2005*, Tsukuba, Japan, Oct 9-13, 2005.

[2]T. Goto and T. Hirai, "Chemically vapor deposited Ti3SiC2", *Mater. Res. Bull.*, **22**, 1195-1201 (1987).

[3]W. Jeitschko and H. Nowotny,"Die Kristallstruktur von Ti3SiC2 - Ein Neuer Komplexcarbid-Typ", *Monatash. Chem.*, **98**, 329-337 (1967)

[4]C. J. Gilbert, D. R. Bloyer, M. W. Barsoum, T. El-Raghy, A. P. Tomsia and R. O. Ritchie, "Fatigue-crack growth and fracture properties of coarse and fine-grained Ti3SiC2", *Scripta Mater.*, **42**, 761-767 (2000).

[5]C. Racault, F. Langlais and R. Naslain, "Solid-state synthesis and characterization of the ternary phase Ti3SiC2", *J. Mater. Sc.*, **29**, 3384-3392 (1994).

[6]R. Radahakrishnan, J. J. Williams and M. Akinc, "Synthesis and high-temperature stability of Ti3SiC2", *J. Alloys Compounds*, **285**, 85-88 (1999).

[7]M. W. Barsoum, T. El-Raghy, C. J. Rawn, W. D. Porter, H. Wang, E. A. Payzant and C. R. Hubbard, "Thermal properties of Ti3SiC2", *J. Phys. Chem. Solids*, **60**, 429-439 (1999).

[8]J. F. Li, W. Pan, F. Sato and R. Watanabe, "Mechanical properties of polycrystalline Ti3SiC2 at ambient and elevated temperature", *Acta Mater.*, **49**, 937-945 (2001).

[9]Y. W. Bao and Y. C. Zhou, "Evaluating high-temperature modulus and elastic recovery of Ti3SiC2 and Ti3AlC2 ceramics", *Mat. Lett.*, **57**, 4018-4022 (2003).

[10]Z. F. Zhang, Z. M. Sun and H. Hashimoto, "Deformation and fracture behavior of ternary compound Ti3SiC2 at 25-1300°C", *Mat. Lett.*, **57**, 1295-1299 (2003)

INFLUENCE OF SPECIMEN TYPE AND LOADING CONFIGURATION ON THE FRACTURE STRENGTH OF SiC LAYER IN COATED PARTICLE FUEL

T. S. Byun, S. G. Hong, L. L. Snead, Y. Katoh
Oak Ridge National Laboratory, P.O. Box 2008, MS-6151, Oak Ridge, TN 37831, USA

ABSTRACT

Internal pressurization and diametrical loading techniques were developed to measure the fracture strength of chemically vapor-deposited (CVD) silicon carbide (SiC) coating in nuclear fuel particles. Miniature tubular and hemispherical shell specimens were used for both test methods. In the internal pressurization test, an expansion load was applied to the inner surface of a specimen by use of a compressively loaded elastomeric insert (polyurethane). In the crush test, a diametrical compressive load was applied to the outer surface(s) of a specimen. The test results revealed that the fracture strengths from four test methods obeyed Weibull's two-parameter distribution, and the measured values of the Weibull modulus were consistent for different test methods. The fracture strengths measured by crush test techniques were larger than those by internal pressurization tests. This is because the internal pressurization produces uniform stress distribution while the diametrical loading technique produces severely localized stress distribution, causing the tremendous reduction of effective surface. The test method dependence of fracture strength was explained by the size effect predicted by effective surface.

INTRODUCTION

Recently there has been a significant international movement for developing gas-cooled nuclear reactors for future energy source and the new reactors will use carbon/carbide coated particle fuels [1-3]. In the coated particles, the spherical fuel kernel is coated by a porous pyrolytic carbon layer (buffer layer) and tri-isotropic (TRISO) layers: inner pyrolytic carbon, SiC, and outer pyrolytic carbon layers. Among these layers, the SiC layer with a thickness of about 40 μm is considered as the most important component for the structural integrity of a fuel particle because it sustains most of the internal pressure from fission gas release in the fuel kernel. Since the outer diameter of the SiC coating in TRISO-coated particle fuel is only about 0.9 mm and its thickness is usually less than 50 μm, the testing and evaluation of such a small component has not been well established [4,5]. Moreover, it is known that the fracture strength of a ceramic material shows size effect depending on specimen type and stress distribution [6,7].

In this work, it was attempted to develop the test methods to evaluate fracture strength of the SiC coating in the TRISO-coated particle fuels. The fracture stress was defined as the maximum stress at fracture and was obtained from the fracture load data using analytical solutions or finite element (FE) analysis for stress distribution. The validity of the developed test techniques was checked by analyzing the test and evaluation results. The size effect on the fracture strength was also investigated by comparing the results because each test method introduces different stress distribution in specimen, causing the change of effective surface. This paper summarizes the developed test methods and the results for fracture strength measurements and their statistical behaviors.

EXPERIMENTAL DETAILS

Table 1. Specimen dimensions and statistical parameters for tubular and hemispherical shell SiC specimens.

Material	Testing method	Specimen dimensions (mm)			Sample size	Statistical strength parameters			
		I.R.**	O.R.**	L**		σ_{mean}*** (MPa)	m	σ_0 (MPa)	S_E (mm^2)
T-SiC-A	IP*	0.511	0.611	5.83	21	263	5.0	448	9.64
T-SiC-B		0.447	0.542	5.27	17	239	9.7	312	8.42
T-SiC-B	DL*	0.447	0.542	5.27	15	283	4.9	405	3.82
S-SiC-A	IP	0.372	0.423	-	17	257	4.3	321	1.734
S-SiC-C		0.383	0.453	-	15	187	6.2	222	1.841
S-SiC-A	DL	0.372	0.423	-	27	822	5.8	427	0.014
S-SiC-C		0.383	0.453	-	31	638	7.5	356	0.008

* IP and DL indicate the internal pressurization method and diametrical loading method, respectively

** I.R., O.R., and L are the inner radius, outer radius, and length of a specimen, respectively.

*** σ_{mean} indicates the mean fracture strength of each type of specimen.

Materials

Tubular SiC specimens were obtained from rod-type surrogate fuels in which the pyrolytic carbon and silicon carbide layers were deposited on graphite rods using CVD process. The graphite rods with a diameter of about 0.9 mm and a length of about 6 mm were coated in a fluidized bed using a vertical high temperature furnace. The pyrolytic carbon coating, the inner pyrolytic carbon layer in TRISO coated particle fuel, was deposited by hydrocarbon gases such as propylene (C_3H_6) and acetylene (C_2H_2) mixed with an inert fluidizing gas (Ar) at about 1300 °C. Then the SiC layer was deposited on the pyrolytic carbon layer by decomposition of methyltrichlorosilane (CH_3SiCl_3) in a hydrogen carrier gas (H_2) at about 1500 °C. To extract SiC tubular shells from the coated rods, both ends of the coated cylinders were removed by grinding and the carbon coating and the graphite substrate were burned off by baking the cylinders in air at 700 °C for 4-6 hours. The fabrication process for the hemispherical shell SiC specimens was nearly the same as that for the tubular specimens although spherical zirconium oxide particles were used for the surrogate kernels (substrates).

Three versions of SiC materials, named SiC-A, B, and C, were produced with almost identical processes. These three materials have the same elastic properties, however, their fracture strengths depend on the manufacturing process. Each type of test specimen was denominated with a prefix reflecting its shape and material: T-SiC-A and T-SiC-B for tubular specimens, and S-SiC-A and S-SiC-C for hemispherical shell specimens. The material and dimensional data for each test specimen are given in Table 1.

Test methods

In the internal pressurization method an elastomeric insert (polyurethane) in the form of cylindrical rod is inserted into a specimen and is axially compressed by pistons (flat-ended pins) to internally pressurize the specimen. The elastomeric inserts were made by casting from liquid polyurethane mixtures or by punching from solid polyurethane rod and then cut to appropriate lengths. In the diametrical loading method, on the other hand, a specimen is compressed by diametrical loading on the outer surfaces using two flat-ended loading columns. For both test methods, crosshead speed was about 0.01 mm/sec. In analyzing the test results, the fracture stress was defined as the maximum tensile hoop (circumferential) stress which occurs at the weakest point on the inner surface in the internal pressurization methods and on the inner surface area just below the loading contact in the diametrical loading. Schematic drawings of each test method are presented in Fig. 1.

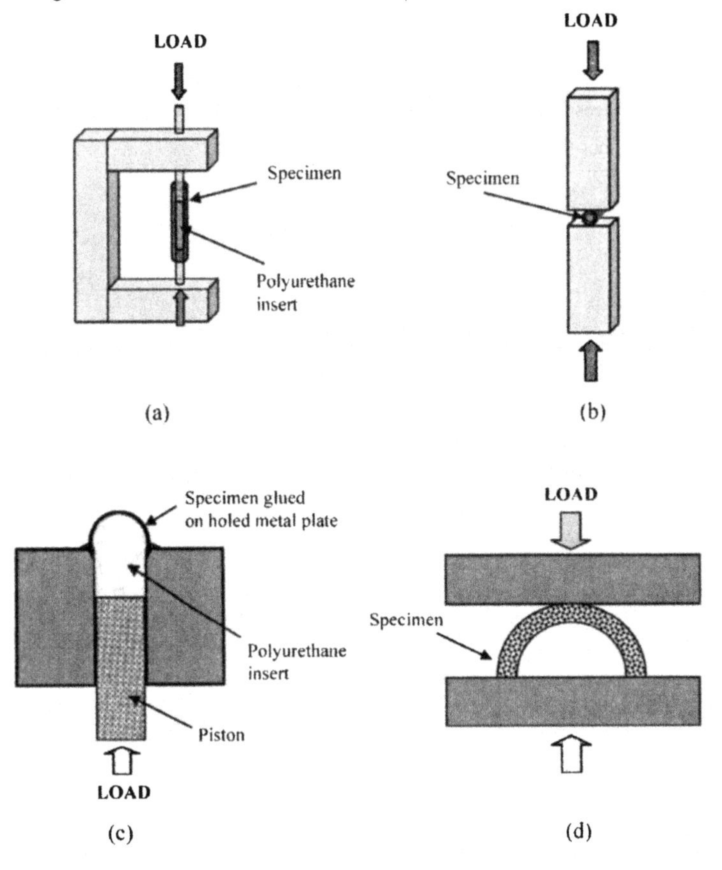

Fig. 1 Schematic drawings of four test methods: (a) internal pressurization test of tubular specimen, (b) crush test of tubular specimen, (c) internal pressurization test of hemispherical shell specimen, and (d) crush test of hemispherical shell specimen.

FRACTURE STRESS CALCULATION

Internal pressurization test with tubular specimen

If a tubular specimen is internally pressurized to a pressure p by compression of an elastomeric insert, the axi-symmetrical hoop stress, σ_θ, is expressed as a function of the radial distance from the centerline of a tubular specimen, r [6-8]:

$$\sigma_\theta(r) = \frac{r_i^2 p}{r_o^2 - r_i^2}\left(1 + \frac{r_o^2}{r^2}\right), \tag{1}$$

where r_i and r_o are the inner and outer radii of a specimen, respectively. Then, the fracture stress, σ_f, which corresponds to the maximum tensile hoop stress when fracture occurs, is given as the stress at the inner surface ($r = r_i$):

$$\sigma_f = \frac{r_o^2 + r_i^2}{r_o^2 - r_i^2} p_f . \tag{2}$$

Crush test with tubular specimen

In the crush test, a tubular specimen is subjected to a diametrical loading. The expression for the stress distribution can be driven from the curved beam theory [9]. If a tube is subjected to a load L applying through the centroidal axis on the outer surface of a tube, the resultant elastic hoop stress is the sum of the stress components caused by an axial load and by a bending load. The bending moment, M, at a point is given by

$$M = \frac{1}{2} LR\left(\frac{2}{\pi(1+z)} - \cos\theta\right), \tag{3}$$

where θ is the angle from the horizontal line,

$$R = \frac{1}{2}(r_i + r_o), \tag{4}$$

$$z = \frac{R}{2c}\ln\left(\frac{R+c}{R-c}\right) - 1, \tag{5}$$

$$c = \frac{1}{2}(r_o - r_i). \tag{6}$$

The hoop stress, σ, at any point (y, θ) is expressed by

$$\sigma = \frac{L\cos\theta}{2a} + \frac{M}{aR}\left(1 + \frac{1}{z}\frac{y}{R+y}\right), \tag{7}$$

where
$a = 2cl$ (the area of cross section),
l = the length of specimen subject to loading,
y = the distance from the mid-thickness position of the wall ($-c \le y \le c$).

The maximum hoop stress occurs at the inner surface of the tube ($y = -c, \theta = 90^o$), inside of the point where load is applied ($y = +c, \theta = 90^o$). Hence, the fracture stress, σ_f, is given by

$$\sigma_f = \frac{L_f}{\pi a(1+z)}\left(1 - \frac{1}{z}\frac{c}{R-c}\right). \tag{8}$$

Internal pressurization test with hemispherical shell specimen

The stress distribution in a spherical container subjected to internal pressure, p, has been derived [10]. This relationship can be also applied to the hemispherical shell specimen, which is internally pressurized to a pressure, p, by compression of an elastomeric insert. The axi-symmetrical hoop stress, σ_θ, in the hemispherical specimen is expressed as a function of the radial distance from the centerline of a hemispherical specimen, r:

$$\sigma_\theta(r) = \frac{r_i^3 p}{r_o^3 - r_i^3}\left(1 + \frac{r_o^3}{2r^3}\right), \tag{9}$$

where r_i and r_o are the inner and outer radii of a specimen, respectively. Then, the fracture stress, σ_f, which corresponds to the maximum tensile hoop stress when fracture occurs, is given as the stress at the inner surface ($r = r_i$):

$$\sigma_f = \frac{(r_o^3 + 2r_i^3)}{2(r_o^3 - r_i^3)} p_f. \tag{10}$$

Crush test with hemispherical shell specimen

No analytical solution for the stress distribution with this loading configuration was found. Thus, the finite element (FE) method was used to calculate the stress distribution in the hemispherical shell specimen and the relation between the applied load and the maximum hoop stress. Compression process of each type of hemispherical shell specimen was simulated and analyzed using a commercial program, ABAQUS. In the analyses, an axi-symmetric model with 4-noded iso-parametric rectangular elements (CAX4R) was used since the configuration of crush test and loading condition are axi-symmetric. The frictions between the SiC shell, upper plate, and lower plate were also taken into account. All component materials in the FE model were assumed to be elastically isotropic. Young's modulus and Poisson ratio used in the calculation were 450 GPa and 0.19 for SiC and 208 GPa and 0.29 for the loading blocks, respectively. The axi-symmetric model for the crush test of the specimen, S-SiC-A, is depicted in Fig. 2. Figure 3 also shows the result of the FE analysis when the compressive load of 12 N is applied.

RESULTS AND DISCUSSION

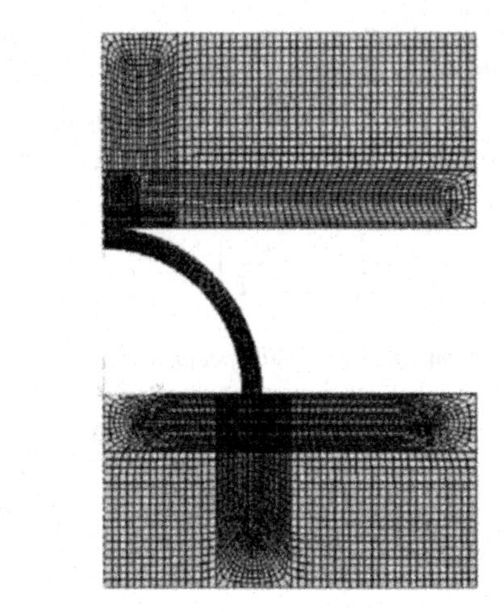

Fig. 2 Axi-symmetric FE model for the crush test of a hemispherical shell specimen.

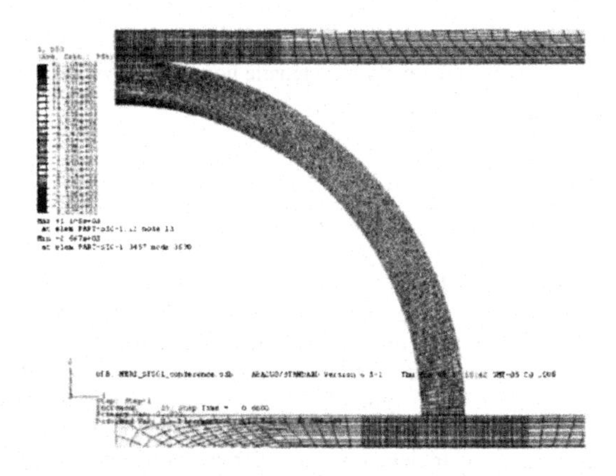

Fig. 3 Stress distribution in the hemispherical shell S-SiC-A specimen during the crush test, where the applied load is -12 N.

It is known that the ceramic materials usually exhibit wide variability and size dependence of strength among seemingly identical specimens. Therefore, the statistical approach, such as the Weibull distribution analysis, has been developed to describe the statistical behavior of the strength of ceramic materials and has been successful in providing reasonable predictions. Weibull statistics is based on a 'weakest link theory', which means that the most serious flaw in the specimen will control its fracture, like a chain breaking by a failure of the weakest link. The most serious flaw is not necessarily the largest one because its severity also relies on its location and orientation related to the stress field. In other words, the flaw subjected to the highest stress intensity factor will be controlling the failure. As will be mentioned later, the inner surfaces of the tubular and hemispherical shell specimens are considered to be the most probable crack initiation locations. Using Weibull's two-parameter distribution, the cumulative probability of failure, P, at or below a stress, σ, is presented by [11,12]:

$$P = 1 - \exp[-\iint_S \left(\frac{\sigma}{\sigma_0}\right)^m dA] = 1 - \exp[-S_E \cdot \left(\frac{\sigma_{max}}{\sigma_0}\right)^m], \tag{11}$$

$$S_E = \iint_S \left(\frac{\sigma}{\sigma_{max}}\right)^m dA, \tag{12}$$

where m, σ_0, and S_E are the Weibull modulus, the scale parameter, and the effective surface, respectively. The Weibull modulus m, also called the shape parameter, represents the scatter in the fracture strength. A higher value of m indicates a lower dispersion of the fracture strength. The scale parameter σ_0 corresponding to the fracture strength with a failure probability of 63.2% is closely related to the mean strength of the distribution. S_E represents the surface area of a hypothetical specimen subjected to uniform stress, σ_{max}, over the whole surface area, which has the same probability of fracture as the test specimen stressed at σ_{max}.

By taking the logarithm twice, Eq. (11) can be rewritten in a linear form:

$$\ln\ln\left(\frac{1}{1-P}\right) = m \cdot \ln\sigma_{max} + \ln\left(\frac{S_E}{\sigma_0{}^m}\right). \tag{13}$$

The Weibull modulus and scale parameter can be obtained from the slope and intercept terms in Eq. (13). Since the true value of P_i for each σ_i is not known, a prescribed probability estimator has to be used as the value of P_i. There have been introduced several probability estimators and their merits have been investigated [13,14]. Among those probability estimators, it is shown that the following probability estimator gives a conservative estimation, and therefore it should be the best choice in reliability predictions [13]:

$$P_i = \frac{i}{N+1}, \tag{14}$$

where P_i is the probability of failure for the ith-ranked stress datum and N is the sample size.

The Weibull plots for the fracture strengths from four test methods are depicted in Fig. 4 and the evaluated Weibull parameters are given in Table 1. The results show that the fracture strength measured by each test method obeys Weibull's two-parameter distribution, and the

Weibull moduli of the same material were consistent regardless of the different specimen shape and test method. The consistency of the Weibull moduli for the same material implies that the

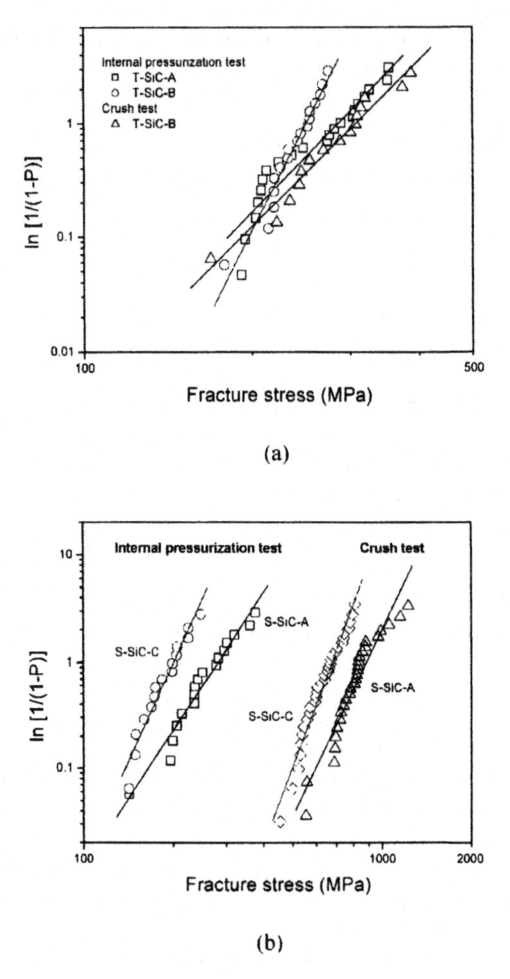

Fig. 4 Weibull plots for the fracture strengths of (a) tubular specimens and (b) hemispherical shell specimens.

developed test methods are reliable because the Weibull modulus is related to the flaw distribution of a material, and thus it is an empirical material constant for the material with the same flaw distribution; the Weibull modulus doesn't vary with the change of specimen geometry and test method [15]. As mentioned earlier the fracture strength distribution of each material used in this study differs from each other because the flaw distribution of each material depends on the manufacturing process. This fact accounts for the difference in the Weibull modulus observed for different materials.

It is noticed that the fracture strengths from the crush tests are larger than those from the internal pressurization tests. For tubular specimens the average fracture strength increases by 18%, while for hemispherical shell specimens a remarkable increase, about 3.2 times for S-SiC-A and 3.4 times for S-SiC-C, was observed. This test method dependence of fracture strength resulted from the change of stress distribution in a specimen, which affects the effective surface of a specimen by the definition, Eq. (12). In the CVD SiC specimens the inner surface is believed to be the most probable crack initiation site not only because maximum stress occurs at the inner surface but also because the inner surface is covered with dimples with a size of a few μm, which originated from the uneven surface of the pyro-carbon substrate. In the crush test of tubular specimens, the stress distribution is highly localized at the inner surface of tubular specimen below the contact with loading column and this highly localized stress is maintained constant along the longitudinal direction. In the crush test of hemispherical shell specimens, on the other hand, the highly localized stress is induced only at the inner surface below the contact with loading column, so the effective surface significantly reduces in comparison with that of the internal pressurization. For a hemispherical shell specimen, the effective surface at fracture under internal pressurization was about 124 times larger than that in the crush test for S-SiC-A and about 230 times for S-SiC-C. When calculating the effective surface of each type of test specimen, multiaxial stress state was taken into account using the principle of independent action model [16,17]:

$$S_E = \int_S \left\{ \left(\frac{\sigma_1}{\sigma_{max}} \right)^m + \left(\frac{\sigma_2}{\sigma_{max}} \right)^m + \left(\frac{\sigma_3}{\sigma_{max}} \right)^m \right\} dA, \tag{15}$$

where σ_1, σ_2, and σ_3 are principal tensile stresses. In the crush test of hemispherical shell specimens, a numerical calculation using the FE analysis results was used since analytic integration for effective surface is impossible.

The effective surface (or volume), in which the effects of multiaxial stress, stress gradient, and failure criterion on the reliability of a material were taken into account, can be used to scale the strength of a ceramic material from one size to another, or from on loading configuration to another [18]. Larger specimens or components are likely to be weaker because of their greater chance to have a larger and more severe flaw. For two specimens having different sizes or loading configuration, the ratio between their mean fracture strengths (or characteristic strengths) can be correlated with the ratio of the effective surface areas:

$$\frac{\sigma_1}{\sigma_2} = \left(\frac{S_{E2}}{S_{E1}} \right)^{1/m}, \tag{16}$$

where σ_1 and σ_2 are the mean (or characteristic) strengths of specimen of type 1 and 2, which may have different sizes and stress distributions, and S_{E1} and S_{E2} are the effective surfaces, and m is the Weibull modulus.

Appling Eq. (16) to the results of tubular specimens, T-SiC-B, the fracture strength by the crush test is predicted to be 11.4% higher than that by the internal pressurization test. As for hemispherical shell specimens, Eq. (16) predicts that the fracture strengths by the crush test are about 2.6 (S-SiC-A) and 2.2 (S-SiC-C) times higher than those by the internal pressurization test. Considering the wide variation in fracture strength data of ceramic materials, this difference between the experimental and predicted data is thought to be acceptable. Therefore the specimen geometry and stress distribution effects on the fracture strength of CVD SiC can be explained with the concept of effective surface. This confirms that the assumption of crack initiation at the inner surface of the CVD SiC layers is reasonable. The statistical parameters for calculation are summarized in Table 1.

CONCLUSIONS

[1] Two types of testing methods, internal pressurization and diametrical loading methods, were developed to evaluate the fracture strength of the SiC layer in TRISO-coated particle fuels and the developed methods were applied to real-sized tubular and hemispherical shell specimens.

[2] It was found that the fracture strengths measured by four test methods could be correlated well by Weibull's two-parameter distribution and the Weibull modulus measured for the same material were consistent.

[3] Higher fracture strengths were measured by the crush tests than by the internal pressurization tests. An extreme case is that the fracture strength of hemispherical shell specimens by crush test was about 3.4 times higher than that by internal pressurization test. This was attributed to the dramatic reduction of effective surface, which was induced by the severe localization of stress distribution under diametrical loading.

ACKNOWLEDGEMENTS

This research was sponsored by the Office of Nuclear Energy Research Initiatives, U.S. Department of Energy, under contract No. DE-AC05-00OR22725 with UT-Battelle, LLC and the Korea Research Foundation Grant funded by Korea Government (MOEHRD, Basic Research Promotion Fund) (No. M01-2005-000-10267-0). The authors give special thanks to Mr. R.A Lowden for producing coated specimens for this study and to Drs. S.J. Zinkle and T. Nozawa for their technical reviews and thoughtful comments.

REFERENCES

[1] P.L. Allen, L.H. Ford, and J.V. Shennan, "Nuclear Fuel Coated Particle development in the Reactor Fuel Element Labortories of the U.K. Atomic Energy Authority," *Nucl. Tech.*, **35**, 246-53 (1977).

[2] G.K. Miller, D.A. Petti, D.J. Varacalle, and J.T. Maki, "Statistical Approach and Benchmarking for Modeling of Multi-dimensional Behavior in TRISO-coated Fuel Particles," *J. Nucl. Mater.*, **317**, 69-82 (2003).

[3] D.A. Petti, J. Buongiorno, J.T. Maki, R.R. Hobbins, and G.K. Miller, "Key Differences in the Fabrication, Irradiation and High Temperature Accident Testing US and German TRISO-

coated Particle Fuel, and Their Implications on Fuel Performances," *Nucl. Eng. Des.*, **222**, 281-97 (2003).

[4] K.E. Gilchrist and J.E. Brocklehurst, "A Technique for Measuring the Strength of High Temperature Reactor Fuel Particle Coatings," *J. Nucl. Mater.*, *43*, 347-50 (1972).

[5] T.S. Byun, E. Lara-Curzio, L.L. Snead, and Y. Katoh, "Miniaturized Fracture Stress Tests for Thin-Walled Tubular SiC Specimens," *Submitted to J. Nucl. Mater.*

[6] O.M. Jadaan, D.L. Shelleman, J.C. Conway, Jr., J.J. Mecholsky, Jr., and R.E. Tressler, "Prediction of the Strength of Ceramic Tubular Components: Part I - Analysis," *JTEVA*, **19**, 181-91 (1991).

[7] D.L. Shelleman, O.M. Jadaan, D.P. Butt, R.E. Tressler, J.R. Hellman, and J.J. Mecholsky, Jr., "High Temperature Tube Burst Test Apparatus," *JTEVA*, **20**, 275-84 (1992).

[8] "Standard Test Method for Tensile Hoop Strength of Continuous Fiber-Reinforced Advanced Ceramic Tubular Specimens at Ambient Temperature Using Elastomeric Insert," Draft for Standard Testing Method in the Annual Book of ASTM Standards proposed by ASTM Subcommittee C28.07 (2002).

[9] F.B. Seely and J.O. Smith, Chapter 6. Curved Flexural Members in Advanced Mechanics of Materials, John Wiley & Sons, Inc., 2nd Ed, August (1961).

[10] S. Timoshenko, Theory of Elasticity, McGraw-Hill, New York (1951).

[11] ASTM Standard, "C1239-00 Standard Practice for Reporting Uniaxial Strength Data and Estimating Weibull Distribution Parameters for Advanced Ceramics," American Society for Testing and Materials, Philadelphia, PA (2003).

[12] M.A. Madjoubi, C. Bousbaa, M. Hamidouche, and N. Bouaouadja, "Weibull Statistical Analysis of the Mechanical Strength of a Glass Eroded by Sand Blasting," *J. Eur. Ceram. Soc.*, **19**, 2957-62 (1999).

[13] B. Bergman, "On the Estimation of the Weibull Modulus," *J. Mater. Sci. Lett.*, **3**, 689-92 (1984).

[14] A. Khalili and K. Kromp, "Statistical Properties of Weibull Estimators," *J. Mater. Sci.*, **26**, 6741-52 (1991).

[15] A. De, S. Jayatilaka, and K. Trutrum, *J. Mater. Sci.*, **12**, 1426 (1977).

[16] R.L. Barnett, P.C. Hermann, J.R. Wingfield, and C.L. Connors, "Fracture of Brittle Materials under Transient Mechanical and Thermal Loading," Tech. Rept. No-TR-66-220, Air Force Flight Dynamics Laboratory, March 1967.

[17] A.M. Freudenthal, "Statistical Approach to Brittle Fracture," in Fracture, Vol. 2, pp. 592-621, Edited by H. Leibovitz. Academic Press, New York, 1968.

[18] D.G.S. Davies, "The Statistical Approach to Engineering Design in Ceramics," *Proc. Br. Ceram. Soc.*, **22**, 429-52 (1973).

INVESTIGATION OF ALUMINIDES AS POTENTIAL MATRIX MATERIALS FOR INERT MATRIX NUCLEAR FUELS

Darrin D. Byler, Kenneth J. McClellan, James A. Valdez
Los Alamos National Laboratory
Material Science & Technology Division
P.O. Box 1663
Los Alamos, NM 87545

Pedro D. Peralta, Kirk Wheeler
Arizona State University
Department of Mechanical & Aerospace Engineering
PO Box 876106
Tempe, AZ 85287-6106

ABSTRACT

New nuclear fuel forms are being sought in an effort to burn down plutonium inventories and for minor actinide transmutation. A study was conducted to screen two potential materials for a new fuel form. In this study, inert matrix fuels (IMFs) were considered as a nuclear fuel type, with particular emphasis on the matrix materials and their compatibility with surrogate oxides/nitrides and cladding materials. In a fuel cycle application, fuel materials need high thermal conductivity, good radiation tolerance, relatively high melting point, ease of fabrication, and suitability for separation. Due to their physical, mechanical and thermal properties, as well as relative ease of fabrication, nickel aluminide (NiAl) and ruthenium aluminide (RuAl) and their solid solutions were considered as potential matrix candidates for IMFs. Such IMFs can be of interest for fast and thermal spectrum applications. This study focused on the ease of fabrication, interaction of molten NiAl and RuAl with the oxides/nitrides, and the compatibility of NiAl and RuAl with the materials in the system, such as oxide/nitrides and typical cladding materials (Zr-4 and HT-9). Results from the experiments indicate limited interaction between the aluminides and cladding materials where inter-diffusion over the 168 hour test period occurred in the worst case a distance of about 11 μm. In light of the results, it was concluded that RuAl and NiAl are promising candidates for IMF materials, warranting further investigation.

INTRODUCTION

Inert matrix fuels (IMF) are of interest for burn down of plutonium inventories using existing commercial power reactors as well as for future fuel cycle applications. These fuels are "inert" in that they do not employ a uranium-based matrix and therefore do not produce actinides. Required matrix properties include high thermal conductivity, good radiation tolerance, relatively high melting points, good fabricability, small neutron capture cross sections relative to Pu, compatibility with cladding and coolant materials, and, in the case of fissile particles dispersed in the matrix, thermodynamic stability in contact with the fissile material and subsequent fission products. Such composite matrix fuels can be of interest for thermal and fast spectrum applications and for recycle or once-through scenarios.

Aluminide intermetallics, especially nickel and ruthenium aluminide, have been identified as interesting matrix candidates due to their combination of high thermal conductivity, relatively high melting points, thermodynamic stability, large database on processing/fabrication, and potential for aqueous reprocessing. Limited irradiation tolerance studies indicate promising performance[1,2]. Solid solutions of NiAl and RuAl can also potentially be used to tailor the thermo-physical properties of the matrix to better match that of the fissile constituent in a composite/dispersion fuel. The compatibility of these aluminide intermetallics with relevant oxide, nitride and cladding materials are not yet established, therefore a study has been carried out to assess the fabricability, and compatibility of the aluminides with surrogate oxides and nitrides as well as typical cladding materials, i.e., HT-9 (ferritic/martensitic stainless steel) and Zr-4 (zircaloy-4). This study included synthesis of materials from the melt, molten interaction experiments of the aluminides with oxides and nitrides, and contact couple compatibility tests at 500 °C and 800 °C. A review of the properties of NiAl and RuAl vis-à-vis the requirement for nuclear fuels will be offered first. Then, a description and discussion will be given of the experimental work to check the compatibility of these aluminides with oxide and nitride surrogate fuel forms, as well as typical cladding materials. Finally, some conclusions regarding the viability of NiAl and RuAl as fuel matrices will be drawn.

Aluminide Properties

Both NiAl and RuAl exhibit the B2 ordered intermetallic phase with the CsCl crystal structure. NiAl has been investigated due to its wide range of physical and mechanical properties and for applications ranging from turbine blades to electronics. NiAl exhibits characteristics that make it desirable as an inert matrix fuel material, such as its thermal conductivity, creep resistance, coefficient of thermal expansion, and melting point, which are tabulated in Table 1 along with the available values for RuAl for comparison purposes. Another consideration when evaluating IMF matrix materials is that of the thermo-physical property requirements of the fissile constituent in a composite/dispersion fuel. It may be possible by forming solid solutions of NiAl and RuAl to better tailor the matrix properties such as coefficient of thermal expansion, fracture toughness, and thermal conductivity to those of the fissile material. Work has been conducted to evaluate the phase equilibria between NiAl and RuAl in the Ni-Al-Ru system, which has lead to a better understanding of the ternary compound.[3] With this and further research, properties of a solid solution between NiAl and RuAl could be tailored to meet many of the requirements of the matrix material. Also associated with the aluminide properties that are desired for a robust fuel form is that of radiation damage tolerance, as discussed next.

Table I. Physical and mechanical properties of NiAl and RuAl.

Material Properties	NiAl	RuAl
Theoretical Density [g/cm^3]	5.91	7.87
Crystal Structure	CsCl	CsCl
Thermal Conductivity (W/mK)	~80	50-70
Thermal Expansion [x10^{-6}/°C]	11-15	12-15
Young's Modulus (GPa)	204.9-0.041T[K]	--
Poisson's Ratio	0.31	--
Thermal neutron cross section (barns)	2.33	1.42
Heat Capacity (J/mole•K)	<60	--
Creep (Thermal/Irradiation)	5x10^{-5}/sec @1300K, 35MPa, thermal neutrons	--
Fracture Toughness (MPa•m$^{1/2}$)	5-6	--
Yield Strength (MPa)	>100	>300
Melting Point (°C)	1638	2050
References	[3-6]	[7-10]

Radiation Damage Tolerance (Heavy Ions)

Heavy ion irradiation of NiAl has been conducted previously by researchers at LANL and elsewhere [1,2,11] as one approach to assess the damage tolerance of stoichiometric NiAl, which may be similar in nature to stoichiometric RuAl. These studies focused on the radiation damage accumulation and the propensity for amorphization in single crystal NiAl. In the LANL study[1], a monocrystalline NiAl sample was irradiated under cryogenic conditions (~100 K) with 450 keV Xe^{++} ions to fluences of 1x10^{13}, 5x10^{13}, and 1x10^{14} Xe/cm^2.[1] The (110)-oriented monocrystalline NiAl target was tilted 7° from normal incidence with respect to the Xe^{++} ion beam, in order to avoid channeling effects during implantation. Figure 1a shows Rutherford backscattering/ion channeling (RBS/C) spectra obtained from unirradiated NiAl (in a channel-oriented and random geometry with respect to a 2 MeV He^+ beam) and from Xe^{++} ion irradiated NiAl to the fluences described above. The results reveal (1) that single crystal NiAl is not amorphized under cryogenic Xe^{++} ion irradiation conditions, at least to a fluence of 1x10^{14} Xe/cm^2; and (2) the rate of damage accumulation was observed to decrease dramatically with increasing fluence. With regard to the latter finding, the damage remained relatively unchanged between fluences 5x10^{13} and 1x10^{14} Xe/cm^2. In other words, retained damage remained constant even though the ion dose was doubled. This result is very unusual, though it was observed in earlier ion irradiation damage studies on single crystal NiAl by Thomé et al.[2] The LANL results are in good agreement with the results published by Thomé et al., as illustrated in Fig. 1b. However, the reasons for the deceleration of radiation damage accumulation and even the decrease in damage observed by Thomé et al. are not understood at this time. While these results are only specifically valid under the conditions tested, they may be taken as an indication that NiAl has promising radiation tolerance and could be a good inert matrix fuel material if it is otherwise compatible with actinides and cladding materials.

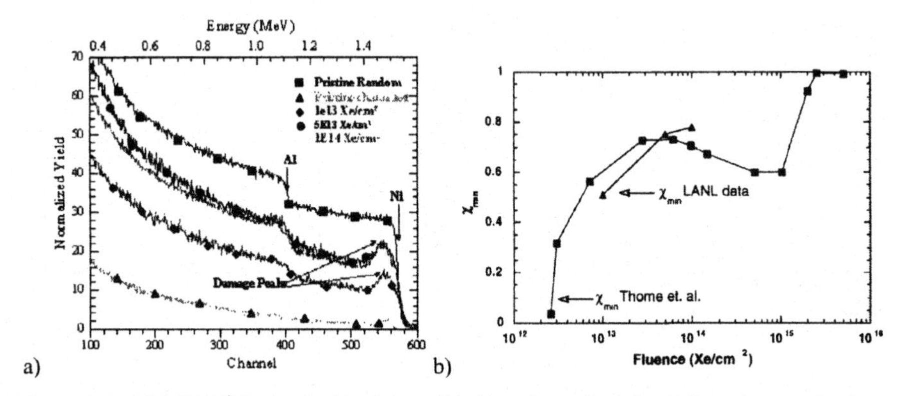

Figure 1 **(a)** 2 MeV He^+ Rutherford backscattering/ ion channeling (RBS/C) spectra obtained from unirradiated monocrystalline NiAl. **(b)** Comparison between the damage accumulation factor, χ_{min}, for RBS/C data obtained by Thomé et al. with that of Sickafus et. al. at LANL.[9,10]

Fabricability of Aluminide Materials

The fabricability of the aluminides is also an important consideration, and due to the melting points of the aluminides, the method of fabrication is limited to only a few processes. Typically, nuclear fuel forms would be tailored in density and composition for the particular reactor environment, which requires the fuel to be fabricated such that it provides for these requirements. One process that has been researched considerably is that of reactive hot isostatic pressing (RHIPing). This technique has been tested for materials such as RuAl with a melting point of ~2050 °C using a hot isostatic press (HIP) with temperatures as low as 600 °C to produce a high-density solid. A related fuel fabrication method for pre-pressed compacts would be combustion synthesis, which is a technique that uses the materials own exothermic (combustion) reaction to synthesize and sinter the powder into a homogenous, high-density material without external pressure being applied during the process. With this process, dimensional control is an issue and post-processing of the fuel would be required to meet specific dimensional tolerances of the fuel stack. Both RHIP and combustion synthesis would rely on a dispersion or composite type pellet where actinide oxide or nitride bearing particles would be mixed with the aluminide matrix material and pressed into a pellet. The pellet would then be sintered by the selected process to provide the desired density/microstructure to allow for fission product accommodation and fission gas release without causing failure of the pellet. While NiAl is considerably easier to process than RuAl and has a multitude of processing routes, not all methods are compatible with fabrication of fuel pellets. However, methods suitable for RuAl should also be suitable for NiAl.[12,13] Also of importance is the compatibility of the aluminides with the materials that they will be in contact with in service, such as the oxides/nitrides and cladding materials. This will be reviewed in the following section.

Contact Compatibility Of Aluminides With Surrogate Oxides and Cladding Materials

The compatibility of the matrix with the fissile constituents in a composite/dispersion fuel has arisen as a key factor in selecting the appropriate matrix material. To facilitate compatibility tests, surrogate oxides and nitrides were selected to indicate possible incompatibilities with matrix materials. Oxides such as alumina, ceria and zirconia were selected as potential surrogates that would have similar responses to oxide fuel particles. Likewise, zirconium nitride was selected to simulate a nitride dispersion fuel material. Aside from the requirements of fuel/matrix compatibility, the matrix material must also be compatible with the available cladding materials that would be used in a reactor.

Considerable emphasis has been placed on the interaction of the matrix materials with the cladding because of the potential fuel-cladding-interaction (FCI) that could lead to failure in service. The importance of which was shown by research conducted at Argonne National Laboratory on the performance of HT-9 (a ferritic/martensitic stainless steel) cladding with a metallic fuel at high temperature to determine the interaction of HT-9 cladding with a metallic fuel. Results indicated that 2 of 15 fuel elements failed and that in the hottest cladding regions, there was a chemical interaction between the fuel and the cladding that resulted in brittle layers in the cladding that were prone to failure.[14] Those experiments had a peak cladding temperature on the order of 660 °C, which was reported as the maximum cladding mid-wall temperature for a fast reactor. Cladding temperatures for light water reactors (LWRs) are closer to 380 °C for the mid-wall maximum temperature, for which they use Zr-4 cladding.[15] The chemical interaction between the pellet and cladding causes a bond that stresses the fuel pin during thermal cycling due to power demands. These stresses arise from a variety of sources including thermal expansion mismatch, radiation swelling, as well as differences in creep rates. Attempts at predicting FCI-induced fuel rod failure have been made, but they tend to treat only the crack propagation aspects of the model and leave the chemical interaction out of the equation.[17] Therefore, it is difficult to know the extent to which a chemical interaction can occur without causing failure, but close attention must be paid to any substantial interaction between materials. Due to these potential interactions, research has been conducted to assess the compatibility of NiAl and RuAl with these cladding materials.

EXPERIMENTAL

Experiments for the preliminary assessment of aluminides for inert matrix fuels were broken into 3 components. 1) Synthesis of aluminides to determine the fabricability of RuAl and to evaluate the microstructure of the materials used for compatibility testing. 2) Liquid-solid compatibility studies where aluminides were arc melted in contact with candidate oxides and nitrides to simulate conditions that might be encountered in service with a coolant failure or accident condition and to screen for any gross interactions that might exclude their use. 3) Solid-solid contact compatibility studies where aluminides were placed in intimate contact with oxides and cladding materials, loaded and held for a total of 168 hours in an inert atmosphere at moderate and high temperatures.

Synthesis of Aluminides

Three compositions of ruthenium aluminide were arc-melted under flowing ultra-high purity (UHP) argon. Samples were melted 3 times each to help ensure homogeneity of the melt. Each of the buttons was then weighed and the density was measured by the immersion density method. Samples were then sliced, ground and polished to a 1 μm finish for optical characterization and SEM analysis.

Liquid-Solid Compatibility

Samples of polycrystalline NiAl and RuAl were melted in contact with candidate oxides and ZrN to determine compatibility with the melt. Samples of the oxides and ZrN were placed as dispersed particles in the aluminide matrix materials and arc melted under flowing UHP argon. The samples were held in the molten state long enough for the aluminides to flow around the dispersed particles. NiAl was melted over alumina, ceria, zirconia and zirconium nitride, while RuAl was melted over zirconia and zirconium nitride due to the higher melting point of the RuAl. Samples were sectioned, ground and polished to reveal the dispersed particles inside the aluminide matrix. Scanning Electron Microscopy (SEM) and optical microscopy were performed on the samples.

Solid-Solid Contact Couple Compatibility

High temperature compatibility experiments were carried out to obtain a preliminary indication of any gross material incompatibilities. Compatibility couples consisted of NiAl/Al_2O_3, RuAl/ZrO_2, NiAl/ZrO_2, RuAl/HT9, NiAl/HT9, RuAl/Zr-4 and NiAl/Zr-4. Experiments were conducted at 500 °C and 800 °C under flowing gettered argon for 168 hours in a mullite tube furnace using a Centorr Furnaces oxygen analyzer to display oxygen level values. Sample coupons were cut, ground and polished into approximately 5 mm x 5 mm squares and sandwiched in stacks that were held in intimate contact using a fixture fabricated from a 304 stainless steel bar with alumina guide pins that kept nickel weights aligned with the sample stacks. Samples were examined using optical microscopy and SEM with Energy Dispersive Spectroscopy (EDS) to analyze for diffusion of species between the aluminides and cladding materials. Samples for the first 800 °C experiment and the 500 °C experiment were analyzed in plan view, while the samples of the second 800 °C experiment were separated and analyzed in plan and cross-sectional views.

RESULTS AND DISCUSSION

Synthesis of Aluminides

Synthesis of melt-derived nickel and ruthenium aluminide samples were conducted to study the starting microstructure of samples used for the compatibility studies. Three RuAl alloy compositions were synthesized, sectioned and polished to reveal grain structure for comparison with published literature. The three alloys had target compositions of $Ru_{49}Al_{51}$, $Ru_{49.5}Al_{50.5}$ and $Ru_{50}Al_{50}$. Analysis of the resultant microstructure of $Ru_{49}Al_{51}$ showed a uniform distribution of the eutectic phase around RuAl grains, while the $Ru_{49.5}Al_{50.5}$ and $Ru_{50}Al_{50}$ samples contained a region in the center that was reduced in or devoid of eutectic composition. It is evident that the arc-melted materials do not exhibit equilibrium microstructures, which may be due to the narrow phase field on the order of 1-2 atomic percent. Analysis of the resultant materials by SEM using EDS provided insight into relative compositions within the ruthenium alloys. Compositional analysis confirmed that the areas of eutectic composition were reduced in aluminum content as discussed in literature.[1,2] Figure 3 illustrates the eutectic phase present within the outer regions of the $Ru_{49.5}Al_{50.5}$ sample button. This corresponds well with micrographs published in the literature.[6]

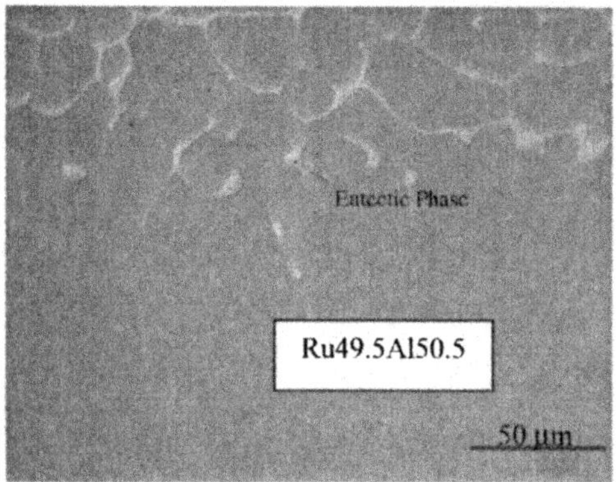

Figure 3. $Ru_{49.5}Al_{50.5}$ sample indicating eutectic composition near outer edge.

Compatibility With Aluminide Melts

Samples of RuAl and NiAl were arc-melted in an inert atmosphere in contact with oxides/nitrides to determine any gross incompatibilities with the molten aluminides. RuAl was melted over ZrO_2 and ZrN samples, while NiAl was melted over Al_2O_3, CeO_2, ZrO_2 and ZrN. Arc-melted samples were sectioned and polished to show the aluminide/ceramic interface. Samples with visible interfaces were first examined by optical microscopy and then by SEM to reveal reaction boundaries, if present.

Samples of NiAl with Al_2O_3, CeO_2, ZrO_2, and ZrN particles were examined for interface reactions. Examination of the NiAl/Al_2O_3 interface indicated that no reaction layers existed, while optical microscopy of the NiAl/CeO_2 interface indicated a large reaction layer shown in figure 4a, which was confirmed by SEM as shown in figure 4b. The NiAl/ZrN sample showed a dark region under the optical microscope (figure 5a) and under the SEM. A eutectic region formed at the boundary of the ZrN and NiAl as shown in figure 5b. Examination of the NiAl/ZrO_2 sample showed similar results to the RuAl/ZrO_2, in that the oxide reduced and incorporated into the matrix.

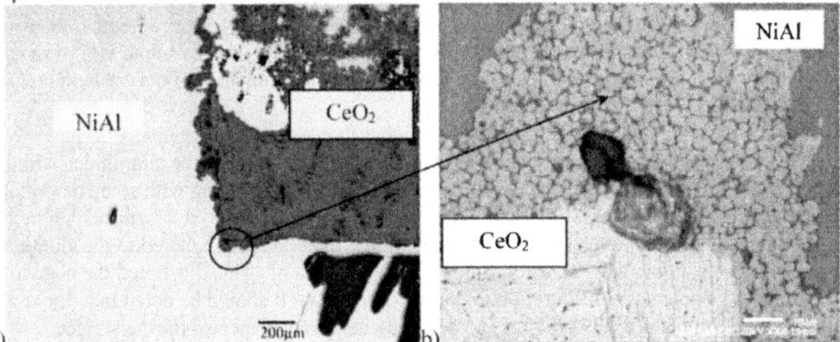

Figure 4 (a) Optical micrograph of CeO_2 inclusion in NiAl matrix and (b) SEM image of structure formed at boundary of NiAl and CeO_2.

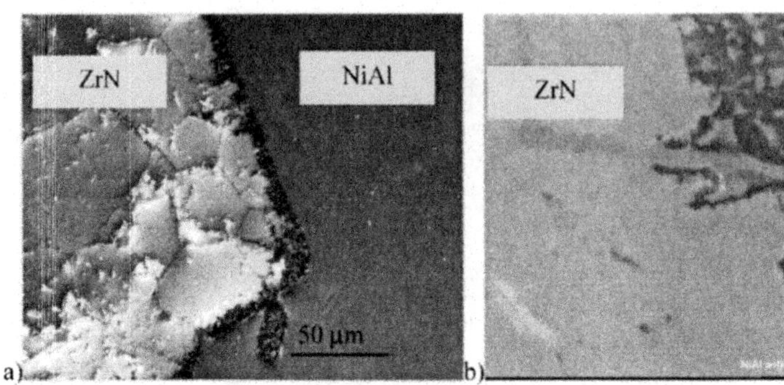

Figure 5 (a) Optical micrograph of NiAl/ZrN interface illustrating reaction layer, and (b) SEM image of potential eutectic region.

Partially based on these results, ZrN and CeO_2 coupons were not included in the contact compatibility studies. Besides the potential issue of reaction of aluminides with nitrides, it should also be noted that little if any fuel performance benefit would be expected in a nitride/aluminide fuel. However, substantial enhancement of thermal conductivity would be expected for oxide fuel particles dispersed in an aluminide matrix. Accordingly, the compatibility couples were focused on compatibility of the aluminides with surrogate oxides and cladding materials.

Solid-Solid Contact Couple Compatibility

Contact couple compatibility studies were designed based on the need to assess the interaction of aluminides with oxides and typical cladding materials. HT-9 and Zr-4 were selected as cladding materials representing applications in fast and thermal reactors, respectively. The preliminary experiment to determine gross effects between the aluminides, oxides and cladding was set at 800 °C, which is higher than the typical temperatures a fast reactor cladding would encounter in service. The higher temperature would enhance any potential interactions. Two additional experiments were also designed with one being at 500 °C and the other at 800 °C, which was an experiment to confirm or negate the results obtained from the 800 °C experiment. Sample sets were furnace cooled and removed after approximately 24 hours at temperature to provide time based information on interactions between materials. These sample sets were examined optically, and then returned to the furnace for an additional 144 hours at temperature to provide consistency between experiments.

During the initial experiment at 800 °C, higher than expected oxygen levels were measured in the furnace, on the order of 40-60 ppmv. Light bonding of the aluminides with the cladding materials was noted after the first 24 hours in the furnace, along with some oxidation of the sample sets. The samples were separated, examined and returned to the furnace. Upon removal after the full 168 hours in the furnace, cladding materials had adhered to the aluminides, which is attributed to the large amount of oxide scale formation. This illustrated the need to improve the experimental configuration to maintain low pO_2. It should be noted that due to the limited volume of gas in a fuel pin, low oxygen contents can be expected during service.

Additional coupons were tested at a temperature closer to reactor service conditions. New sets of samples were made and tested at 500 °C for 168 hours under a gettered argon flow that provided a pO_2 of approximately 10^{-12} ppmv for the duration of the experiment. The samples were again examined at 24 hours and 168 hours, showing now signs of bonding between the aluminides and cladding materials and only slight discoloration in the form of multi-colored rings. Materials were examined optically and using SEM with EDS capabilities at Arizona State University. Figure 7 illustrates a typical sample set from the 500 °C experiment after 168 hours, where contact areas between the different materials can be clearly seen. There was a slight indication of diffusion between NiAl and Zr-4 in the line scan as shown in figure 8a and 8b. The location of the increase in zirconium is indicated by a (+) in figure 8a and on the line scan in figure 8b.

Figure 7 Image of NiAl/Zr-4 couple after removal at 168 hours showing banded area around the outer edges of the samples.

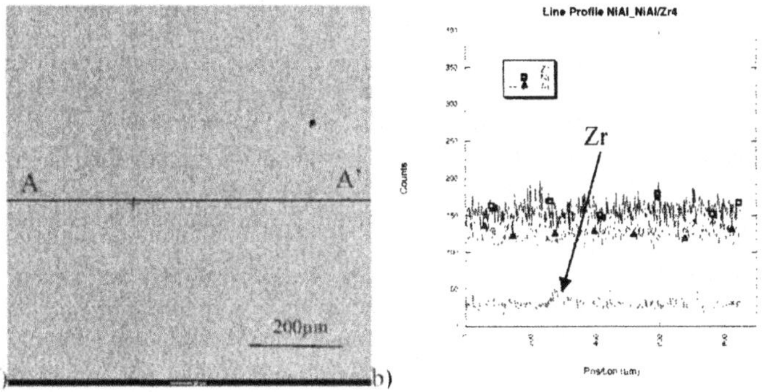

Figure 8 (a) Backscatter image of the plan view of a NiAl sample from the NiAl/Zr-4 contact couple, (b) Line profile of plan view of sample.

Following the 500 °C experiment, an additional confirmatory experiment was conducted at 800 °C to determine whether the results of the first experiment were accurate. Oxygen levels were kept as low as possible during the experiment to reduce the chance of oxide scale formation that could alter the results. Preliminary observations showed bonding of the cladding materials after 24 hours in the system. Unlike the first experiment. the couples were not separated; therfore, no qualitative indication of bond strength was made. Samples were removed after 168 hours for complete optical and SEM-EDS characterization. Initial observations showed that the NiAl/Zr-4, RuAl/Zr-4, and RuAl/HT-9 had remained bonded throughout the experiment. These materials were examined optically and then under the SEM using EDS to characterize the surfaces of the samples. Analysis of plan view samples showed diffusion of species from one coupon to the other and line scan analysis on cross-section samples illustrates this well. The most pronounced interaction occurred between the NiAl and Zr-4. This is shown in figure 9a with the cross-section of the NiAl/Zr-4 interface and the associated line profile shown in figure 9b.

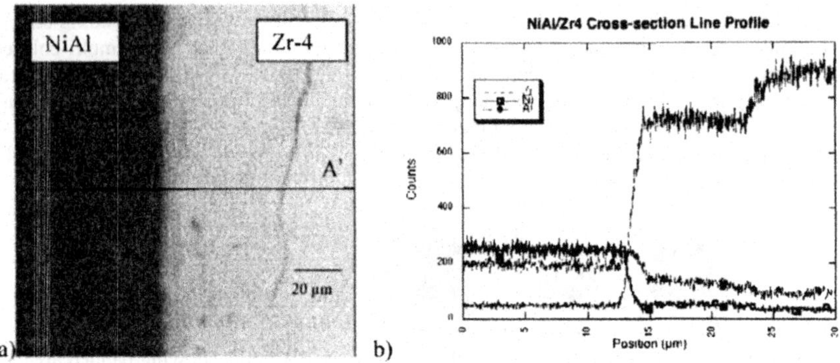

Figure 9 (a) Backscatter image of NiAl/Zr-4 cross-section. (b) EDS line profile indicating elemental change through interface over ~11 μm.

Analysis of the line profile and image indicate that there is an enrichment of aluminum in the Zr-4 at the interface with the NiAl and that the reaction region extended to ~11 μm over the course of the combined 168 hour experiment. This resulted in a relatively strong bond between the two materials. Analysis of the RuAl/HT-9 indicated that there was also an interaction area on the order of 3 μm where the Ru and Al appear to have diffused into the HT-9. Examination of the image and spectra from the RuAl/Zr-4 also showed the appearance of a small amount of diffusion between the Ru and Al into the Zr-4.

Overall, no extensive interaction between the materials was observed for the 500 °C experiments, and no clear, application-limiting interaction between materials in the 800 °C experiments was observed. The maximum observed extent of diffusion was limited to approximately 11μm over the entire period of 168 hours at temperature and the materials that attained the highest diffusion were those typically intended for thermal reactors with temperatures much lower than the test temperature.

CONCLUSION

The preliminary assessment for the compatibility of the aluminides as matrix materials in IMFs indicates that due to NiAl's and RuAl's high melting point and the reactive nature of the melt, solid state synthesis methods should be considered, for example, combustion synthesis should be suitable for fabrication of dense aluminide matrix fuels containing dispersed oxide fuel particles. Additionally, there are no observed incompatibilities between the aluminides and the considered oxides and claddings materials at 500 °C, even though some cross-diffusion was seen between NiAl and Zr-4. A larger interaction between the aluminides and the cladding materials occurred at 800 °C, where the greatest extent of diffusion was only about 11 μm after 168 hours. Given the fact that in-pile radiation has the potential to increase the diffusion kinetics of these materials, the extent of the effect is not known and will require actual testing with actinide bearing materials. However, these preliminary results indicate that the aluminides are worthy of further examination to determine whether they are sufficiently compatible with oxide fuels and cladding materials.

ACKNOWLEDGEMENTS

This research was funded through the Advanced Fuel Cycle Initiative under aegis of the U. S. Department of Energy and administered by the University of California.

REFERENCES

[1]Sickafus, K., Valdez, J.A., Egeland, G., Radiation Effects Studies in Transmutation Fuel Matrix/Diluent Materials, *AFCI Quarterly Report*, 2003.

[2]L. Thomé, C. Jaouen, J. P. 'Rivière and J. Delafond, "Phase transformations in ion irradiated NiAl and FeAl",*Nucl. Instr. and Meth.* B **19/20** 554-558, (1987).

[3]S. Chakrovorty, D. West, "Phase equilibria between NiAl and RuAl in the Ni-Al-Ru system". *Scripta Metallurgica* **19**, 1355-60 (1985).

[4]D. Miracle, " The physical and mechanical properties of NiAl", *Acta Metallurgica et Materialia* **41**, 649-84 (1993).

[5]R. Noebe, R. Bowman, M. Nathal, "Physical and mechanical properties of the B2 compound NiAl", *International Materials Reviews* **38**, 193-232 (1993).

[6]S. Deevi, V. Sikka, C. Liu, "Processing, properties and applications of nickel and iron aluminides", *Progress in Materials Science* **42**, 177-92 (1997).

[7]S. Anderson, C. Lang, "Thermal conductivity of ruthenium aluminide (RuAl)", *Scripta Materialia* **38**, 493-7 (1998).

[8]Wolff, I., "Toward a better understanding of ruthenium aluminide", *JOM*, **49**(1): p. 34-39 (1997).

[9]B. Tryon, T. Pollock, M. Gigliotti, K. Hemker, "Thermal expansion behavior of ruthenium alumindies", *Scripta Materialia* **50**, 845-848, (2004).

[10]M. Cortie, T. Boniface, "Synthesis and processing of ruthenium aluminide", *Journal of Materials Synthesis and Processing* **4**, 413-28, (1996).

[11] M. Natasi, J.W. Mayer, "Thermodynamics and kinetics of phase transformations induced by ion irradiation", *Mat. Sci. Reports*, **6**, pp.1-51, (1991)

[12] I.M. Wolff, "Synthesis of RuAl by Reactive Powder Processing", *Met. And Mat. Trans A*, 27A, 3688-3698, (1996).

[13] M.B. Cortie, T.D. Boniface, "Synthesis and processing of ruthenium aluminide", *J. Mat. Synth. And Proc.*, 4(6), 413-427, (1996).

[14] R.G Pahl, C.E. Lahm, S.L. Hayes, "Performance of HT9 clad metallic fuel at high temperature", *J. Nucl. Mat.*, **204**, pp.141-7, (1993).

[15] D.R. Olander, "Fundamental Aspects of Nuclear Reactor Fuel Elements", Technical Information Center, Office of Public Affairs Energy Res. and Dev. Admin., p.118. (1976)

[16] S.P. Walker, Yu, A., Fenner, R.T., "Pellet-clad mechanical interaction: pellet-clad bond failure and strain relief", *Nucl. Eng. Des.*,**138**, 403-8, (1992).

[17] L.O. Jernkvist, "A model for predicting pellet-cladding interaction-induced fuel rod failure", *Nucl. Eng. and Design*, 156, 393-9, (1995).

FLUIDISED BED CHEMICAL VAPOUR DEPOSITION OF PYROLYTIC CARBON

E. López Honorato, P. Xiao*
School of Materials, University of Manchester.
Grosvenor Street, Manchester M1 7HS, UK.

G. Marsh, T. Abram
Nexia Solutions Ltd.
Springfields PR4 0XJ, UK.

ABSTRACT

Pyrolytic carbon was deposited from acetylene and a mixture of acetylene/propylene on alumina particles at temperatures ranging from 1250 to 1450°C. The depositions were performed in a fluidised bed chemical vapour deposition (FBCVD) reactor at atmospheric pressure. Different techniques such as scanning electron microscopy (SEM), Raman spectroscopy and transmission electron microscopy (TEM) were applied to study the effect of deposition temperature, precursor concentration and flow rate. Difference in density and microstructure resulted from different types of products formed during the decomposition of the gas precursor and also depended whether formation of pyrolytic carbon occurred at the surface of the coated particles or in gases.

INTRODUCTION

Pyrolytic carbon (PyC) can be defined as the carbon material obtained from the decomposition of gaseous hydrocarbon compounds in an activated environment (heat, plasma)[1]. Among other applications, PyC has been used in the nuclear industry as coating layers for fuel particles in advanced high temperature gas cooled reactors (HTR)[2, 3]. This type of nuclear fuel, also known as TRISO, serves as a containment vessel to retain most of the fission products within the particle. They are formed by a fissile kernel, typically UO_2, surrounded by four ceramic layers, three of them made of PyC. The first layer, a 95 μm low density PyC, absorbs the recoil damage and provides void space to retain any gaseous fission products released from the kernel. The remaining PyC layers, typically two 40 μm thicknesses of high density, act as a barrier to retain some of the fission products, sealing off the kernel and protecting the intermediate SiC layer from mechanical damage [4, 5].

Due to its importance, extensive research was carried out during the 60's and 70's, trying to understand and correlate the deposition conditions of fluidised bed chemical vapour deposited PyC, to its mechanical properties and microstructure [6,7,8]. Despite the great progress achieved during this time, little further research was carried out after the HTR projects around the world were cancelled. Even though work on PyC materials has continued, most of the current experiments have been done using horizontal or vertical hot wall Chemical Vapour Deposition (CVD) reactors [9,10,11]. The difference in the CVD configurations may lead to a large difference in deposition conditions and results. Due to this reason, a real comparison among these experiments is not always possible.

As a result of the renewed interest in developing HTR technology, a fluidised bed CVD equipment was designed and built for this specific purpose. In the present study we describe the effect of temperature, precursor concentration and flow rate on the properties of PyC, establishing possible explanations about the influence of each variable.

EXPERIMENTAL

The fluidised bed CVD reactor used in these experiments consists of a graphite tube (200 mm length, 35 mm internal diameter) heated by a resistance graphite element and a multi-hole nozzle as gas distributor. About 30 g of fully sintered alumina microspheres (500μm diameter) were fluidised with argon in a spouted mode. PyC was deposited at atmospheric pressure with temperatures in the range of 1250–1450°C. Acetylene (40-70% v/v concentration) and a mixture of acetylene/propylene (1:1.5; 20-40% v/v concentration) were used as precursors. Three different flow rates (3, 4 and 4.5 L/min) were used during PyC deposition with acetylene.
Microstructural features were characterised using a scanning electron microscope (field emission gun Philips XL30 FEG-SEM) and a Renishaw Ramascope with a 514 nm Ar-ion laser source. Single spot measurements were performed in polished cross sections using a 50X magnifying objective lens, which focused the beam to a spot size of around 1μm in diameter. Measurements were collected over the cross section at 10μ away from the alumina surface. TEM examination was performed in a Phillips CM200. Samples were ground, suspended in acetone and placed on a cupper grid. Densities were obtained by using the Archimedes method in ethanol. Porosimetry measurements were carried out using a PoreMaster porosimeter (Quatachrome Instruments).

RESULTS

Density

Fig 1 shows the density of PyC produced from both types of precursor. The density decreased with increasing temperature and to a certain extent with increasing acetylene (Ac) concentration. A slightly different pattern was observed for the mixture Acetylene/Propylene (Ac/Prop) precursor where the density results are similar from 1350 to 1450°C, regardless the precursor concentration. In addition, the effect of precursor on density is very small. The temperature effect on density is less significant than that using Ac precursor.

SEM observations of PyC from acetylene confirmed these variations. Porosity clearly increased with increase in both pore size and pore number as temperature increased. While PyC at 1250°C shows no apparent pores, pores with size of ~5μm were observed in PyC produced at 1450°C (Fig. 2). These changes in porosity were also confirmed by measurements using porosimetry analysis. Fig. 3 shows pore volume as function of pore size of PyC produced using 50% Ac as precursor. The porosity was generated from larger pore size and volume as deposition temperature increased, even showing the presence of pores of around 11μm at 1450°C. Similar effects were observed when the acetylene concentration increased. At 1400°C, 40% acetylene gave a relatively porous PyC. However, when 70% concentration was used, bigger pores were obtained reaching sizes of around 15μm diameter. In addition to these differences, surface roughness changed from a partially flat and compact surface at 1250°C, to a quite rough and porous at 1450°C.

The effect of temperature on PyC density was also evident during SEM observation. With the mixture acetylene/propylene as precursor, very little porosity was noticed at 1300°C. In contrast, far more porosity was generated at 1450°C.

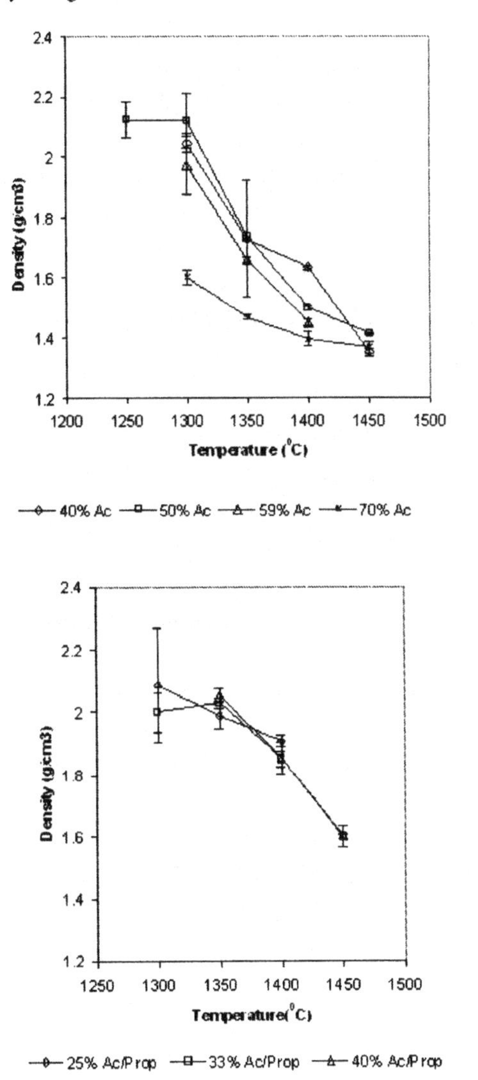

Fig. 1. Effect of temperature and concentration on the density of PyC.

The influence of flow rate on the deposition of PyC was also studied. SEM images indicate that porosity and surface roughness increased with flow rate (Fig. 4). Density measurements on 50% Ac at 1400°C supported this result by showing a reduction of the density from 1.5 g/cm^3 to 1.4 g/cm^3 as flow rate increased from 3 L/min to 4.5 L/min (Fig. 5).

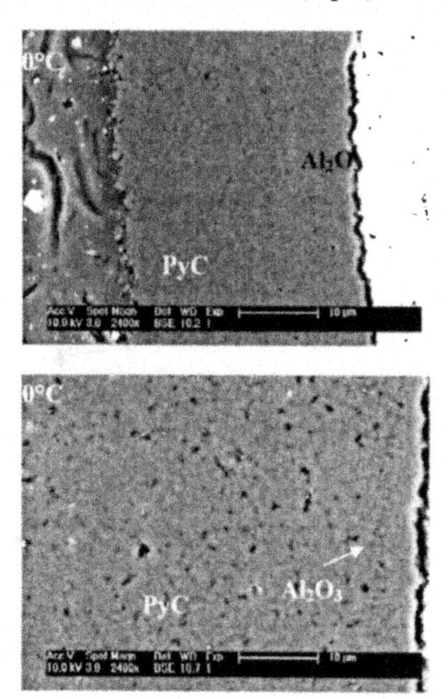

Fig. 2. SEM micrographs showing the effect of temperature on PyC (50% Ac).

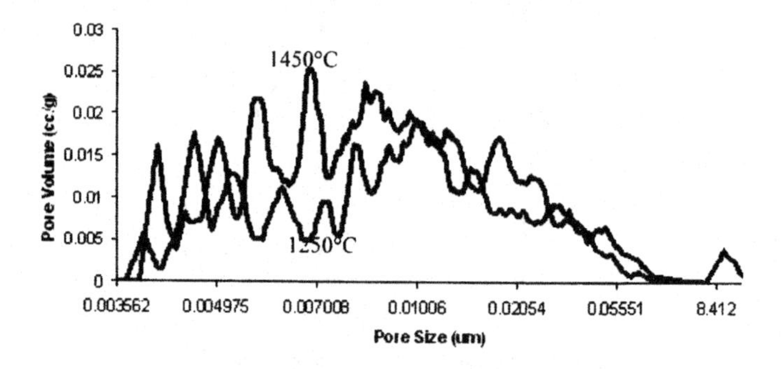

Fig. 3. Porosimetry analysis of PyC at 50% Ac.

Crystal Size and Crystal Order

According to previous studies[13,14], Raman spectroscopy can provide information about the crystal size and crystalline order in carbon materials. The main features of the Raman spectra of carbon are two first-order bands also known as D

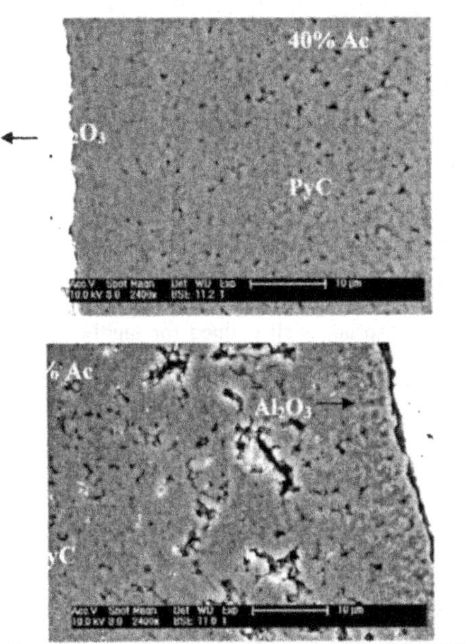

Fig. 4. Effect of concentration on PyC (1400°C).

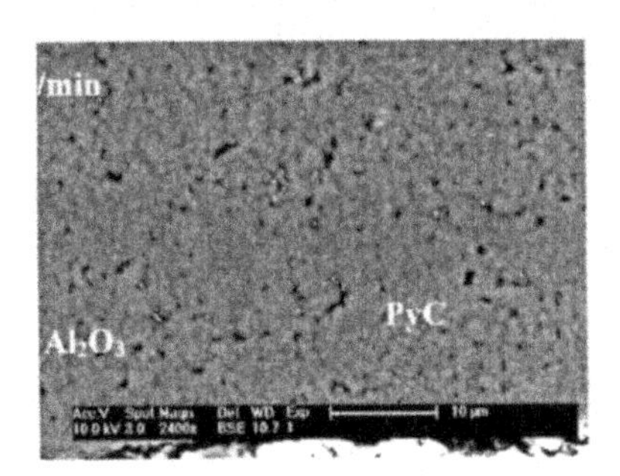

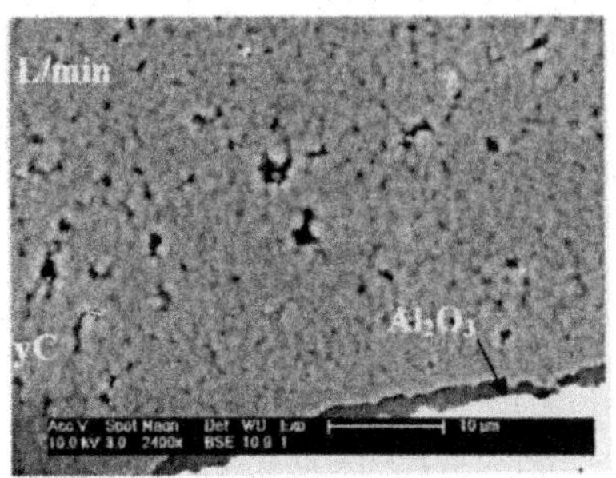

Fig. 5. Effect of flow rate on PyC deposition (50% Ac; 1400°C).

(1330-1360 cm^{-1}) and G (1580-1600 cm^{-1}) bands. The D band is associated to the presence of disorder arising from the crystal boundaries of larger crystallites. The D band grows in intensity with increasing degree of disorder. The G band is attributed to an in-plane stretching vibration in the basal plane of carbon. Line broadening of the G band is also related to the disorder within the carbon sheet. The G peak is sharp and well defined for highly oriented pyrolytic carbon and broadens considerably in disordered carbon films.

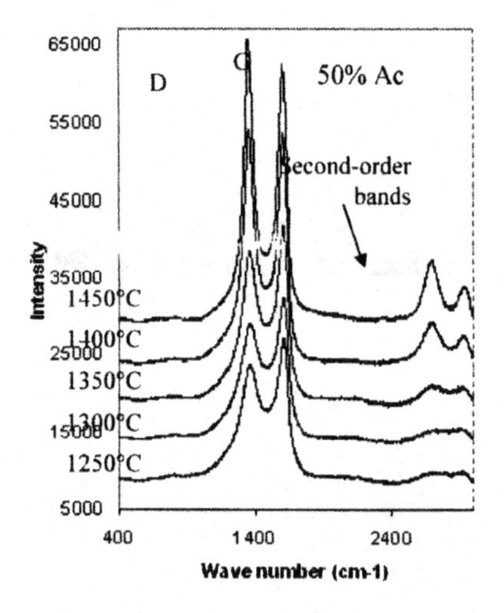

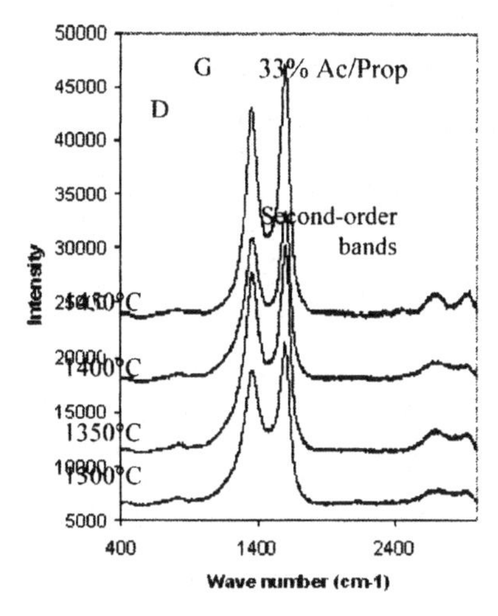

Fig. 6. Effect of temperature on the Raman spectra of PyC.

The intensity ratio of the D and G bands is inversely proportional to the in-plane crystal size L_α[12]:

$$L_\alpha\,(\text{Å}) = 44(I_D/I_G)^{-1} \qquad (1)$$

In addition to the first-order bands, which are related to the structural order within the carbon sheets, second-order bands (around 2700 cm^{-1}) can be present in the carbon spectra. These bands are associated to the stacking disorder along the c-axis or three dimensional ordering [13, 14].

In Fig. 6, acetylene-derived PyC exhibits a considerable increase in D band intensity together with deposition temperature. This change is not apparent for the PyC produced from the Ac/Prop mixture. The increase in D band suggests that the in-plane disorder is higher at elevated deposition temperature in the PyC produced from Ac, but not the PyC from the Ac/Prop mixture. In contrast, based on the variation of the second-order bands (Fig. 6), the three dimensional ordering appeared to increase with increase in temperature for both types of precursors. This tendency was also observed with increase in precursor concentration (Fig. 7) and flow rate (Fig. 8). However, the change in precursor concentration (Fig. 7) and flow rate (Fig. 8) gave less change on G and D band than that by temperature change.

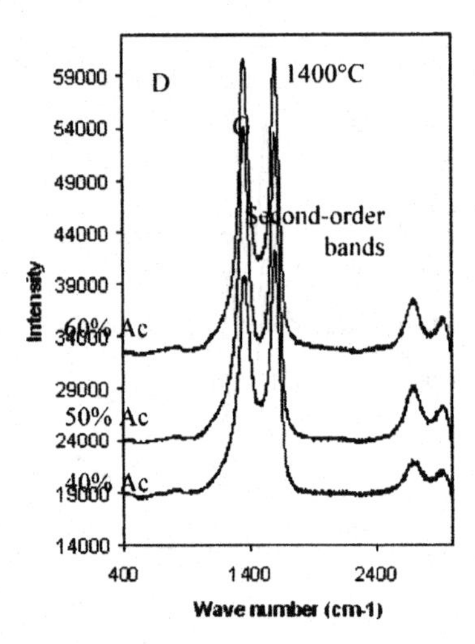

Fig. 7. Effect of precursor concentration on the Raman spectra of PyC.

Assuming that disorder is mainly caused by the presence of grain boundaries, we present the crystallite size as function of temperature and Ac concentration in Fig. 9, based on Fig. 6 and equation 1. The crystal size decreased with increase in the deposition temperature and Ac concentration (Fig. 9). Crystal sizes ranged from 54Å to 40Å as the deposition temperature was changed from 1250°C to 1450°C for PyC produced from 50% Ac. A difference of around 5Å was measured between PyC crystal sizes obtained at 40%Ac and the other two concentrations. The concentrations of 50 and 59% Ac produced almost the same crystallite size with the different deposition temperatures. However, the PyC from Ac/Prop mixture shows slightly increasing crystal sizes with increase temperatures. In addition, crystallite sizes decreased with increase in the Ac/Prop mixture concentration (Fig. 10). Clearly the addition of propylene produced some changes in the deposition process which resulted not only in higher densities but also higher crystal sizes and crystalline order. Furthermore, higher flow rates modified the crystal size by reducing them from 47Å to 42Å as the flow rate increases from 3 to 4.5 L/min.

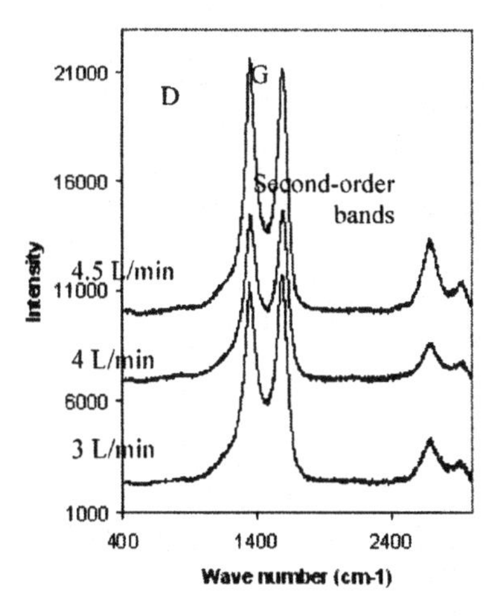

Fig. 8. Influence of flow rate on the Raman spectra of PyC (50% Ac, 1400°C).

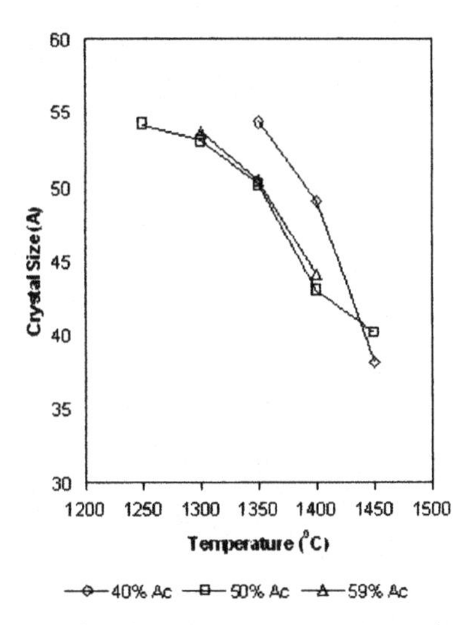

Fig. 9. Crystal size as function of temperature and acetylene concentration.

DISCUSSION

Different authors have tried to correlate the deposition conditions with PyC properties, however up to now there is no single theory capable of explaining the deposition mechanism and the results observed. One of the difficulties in trying to produce a single theory is the complexity of the system and that in general, each research group uses different set of conditions and different types of CVD facilities.

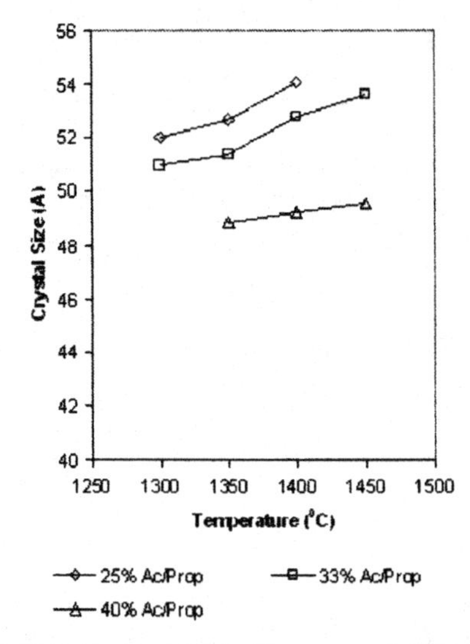

Fig. 10. Effect of temperature and Ac/Prop concentration on crystal size.

It general it has been accepted that hydrocarbons decompose through a series of steps to form pyrolytic carbon with different degrees of crystalline order. The decomposition of these hydrocarbons, known as gas phase maturation, is as follows: (i) cracking of the initial molecules into more reactive and lighter gaseous species, (ii) recombination of these species up to the synthesis of linear and light aromatic compounds, and (iii) evolution of these aromatic compounds towards higher molecular weights (Polycyclic Aromatic Hydrocarbon or PAH)[15]. Several experiments have proved that this maturation is affected by different parameters such as temperature, residence time and type of hydrocarbon[16, 17]. Gas maturation is favoured by increasing T, pressure and residence time[18].

Taking into consideration previous results, it was concluded that low textured carbon (low anisotropy and low density) resulted from the condensation of large PAH molecules[1, 18]. Conversely, high textured carbon (high anisotropy and high density) occurred from the deposition of linear and small aromatic compounds. In addition to these results, a condensation and growth mechanism was proposed[19].

During our experiments we observed that PyC density decreased as well as the coating rate increased with increasing deposition temperature. precursor concentration and flow rate. The high density obtained at around 1250 and 1300°C is partly due to the deposition of high texture carbon which can be well compacted and reduce the amount of porosity and defects.

As the temperature was increased the density of PyC was reduced for both type of precursors studied. This reduction in density may result from two different processes. Firstly, as the temperature is increased, the formation and supply rate of growth species to the growth sites is improved due to an increased surface mobility. In other words, more 5 member rings (PAHs) formed part of the PyC. As mentioned before the addition of such species result in the formation of large nanoporosity. TEM images showed that at high temperatures (1450°C) carbon layers are highly disordered, including a large amount of curvatures. These curvatures produced the amount of nanoporosity previously observed by porosimetry. In contrast, when high density PyC was analysed, carbon layers were parallel to each other and almost no curvatures were observed (Fig. 11).

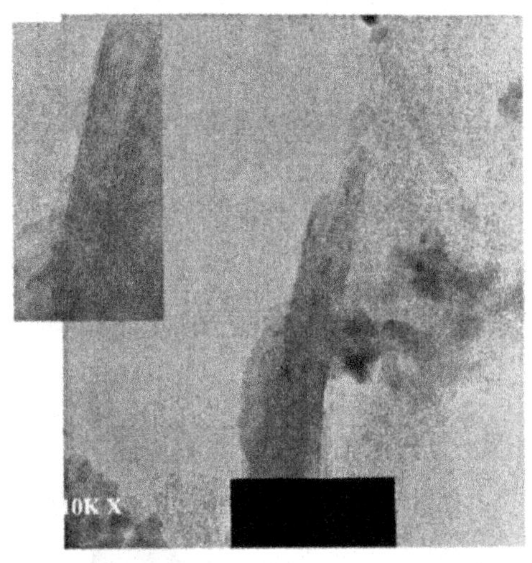

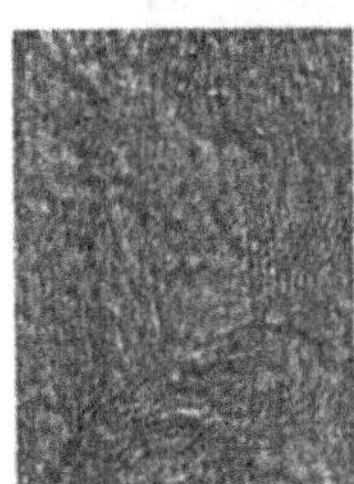

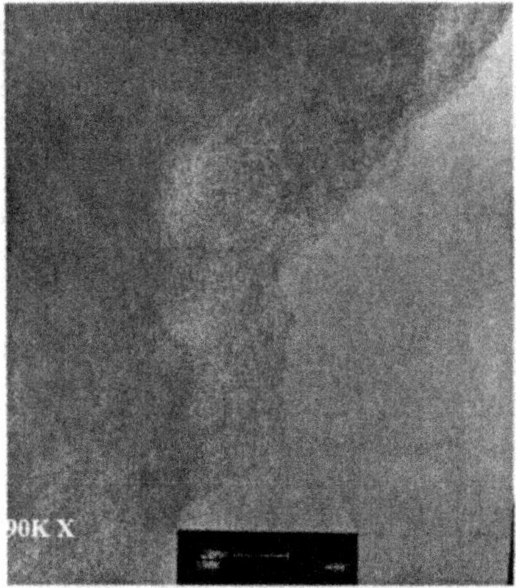

Fig. 12. TEM images of PyC obtained at 1250°C(a) and 1450°C(b).

In addition, it is well established that increase in temperature leads to increase in the rate of homogeneous reactions in the gas phase. These reactions induced the formation of soot, which in turn is deposited on the surface of the material. The SEM images showed the presence of soot in all the low density coatings (Fig. 12). More interestingly is the fact that most of the large porosity seems to be produced by the presence of soot. As the soot is deposited it interrupted the growth of PyC leading to the formation of pores.

Fig. 12. Soot deposits on PyC

Similarly, as the flow rate is increased the residence time is reduced, leading to the formation of more small aromatic or linear molecules and lower densities. Eroglu and Gallois suggested that the increase in growth rate could be explained by a mass-transport theory where the boundary layer thickness is reduced as the gas velocity is increased[20]. This reduction in the boundary layer thickness together with a higher mobility of the reactive species results in high coating rate.

The precursor concentration has a similar effect. Considering that the actual amount of hydrocarbon in the reaction zone is increased, the amount of intermediate species available for

PyC formation is higher, producing a rise in the coating rate. As the coating rate increases, the probability of misalignments within the PyC also increases, due to the lack of time for rearrangement between newborn crystallites.

During the low temperatures examined in these experiments, the amount of homogeneous reactions was reduced, and most of the intermediate species formed at the surface of particles. The surface reaction promoted crystal nucleation, followed by crystal growth when adjacent crystals meet at the grain boundaries. These processes result in larger crystal sizes with better crystal orientation, explaining the behaviour seen in acetylene-derived PyC. When more than one hydrocarbon is added to the reaction zone, a wider range of intermediate species are formed with some of them being more reactive than others. One important difference between acetylene and propylene is that even though acetylene is very reactive and can form pyrolytic carbon directly, propylene (C_3-specie) was found to be not only more reactive but also to form products of higher activity in carbon formation (higher aromatic hydrocarbons)[21,22]. The combination of higher reactivity together with the availability of more molecules capable of producing planar arrays gave different structure of the PyC.

CONCLUSIONS

Density, crystal size and crystal order have been studied as function of three deposition parameters: temperature, precursor concentration and flow rate. The studied properties have been found to be related to the deposition variables in the following manner:

1. Density decreased with increasing deposition temperature, precursor concentration, and flow rate.
2. In-plane crystalline order or/and crystal sizes tended to decrease with increase in temperatures or flow rates on acetylene-derived PyC. The opposite behaviour is observed in acetylene/propylene PyC.
3. Homogeneous or heterogeneous reactions lead to differences in pore size and porosity distribution in PyC.

Based on these results it would be possible to produce the specific types of PyC used in the production of coated fuel particles by the combination of different deposition parameters.

FOOTNOTES

* To whom correspondence should be addressed (ping.xiao@manchester.ac.uk)

ACKNOWLEDGMENTS

The authors would like to thank Nexia Solutions Ltd. for the financial support provided. In addition, they would like to thank CONACYT-México for a PhD grant to E. López Honorato.

REFERENCES

[1]O. Feron, F. Langlais, R. Naslain, and J. Thebault, “On kinetic and microstructural transition in the CVD of pyrocarbon from propane”, Carbon, **37**, 1343-1353 (1999).

[2]R. W. Dayton, J. H. Oxley, and C. W. Townley, “Ceramic Coated Particle Nuclear Fuels”, J. Nucl. Mat., **11**(1), 1-31 (1964).

[3]H. Nabielek, G. Kaiser, H. Huschka, H. Ragoss, M. Wimmers, and W. Theymann. “Fuel for the pebble-bed HTRs”, Nucl. Eng. Des. **78**, 155-166 (1984).

[4]G. H. Lohnert, H. Nabielek, and W. Schenk, "The fuel element of the HTR-module, a prerequisite of an inherently safe reactor", Nucl. Eng. Des., **109**, 257-263 (1988).

[5]R. E. Bullock, "Irradiation performance of experimental fuel particles coated with silicon-alloyed pyrocarbon: a review. J. Nucl. Mat. **113**, 81-100 (1983).

[6]R. J. Bard, H. R. Baxman, J. P. Bertino, and J. A. O'Rourke, "Pyrolytic carbons deposited in fluidised beds at 1200 to 1400°C from various hydrocarbons", Carbon, **6**, 603-616 (1968).

[7]J. C. Bokros, "The structure of pyrolytic carbon deposited in a fluidised bed", Carbon, **3**, 17-29 (1965).

[8]L. H. Ford, N. S. Hibbert, and D. G. Martin, "Recent developments of coatings for GCFR and HTGCR fuel particles and their performance", J. Nucl. Mat. **45**, 139-149 (1972).

[9]Z. J. Hu, W. G. Zhang, K. J. Hüttinger, B. Reznik, and D. Gerthsen, "Influence of pressure, temperature and surface area/volume ratio on the texture of pyrolytic carbon deposited from methane", Carbon, **41**, 749-758 (2003).

[10]G. L. Vignoles, F. Langlais, C. Descamps, A. Mouchon, H. Le Poche, N. Reuge, and N. Bertrand, "CVD and CVI of pyrocarbon from various precursors", Surf. Coat. Tech., **188-189**, 241-249 (2004).

[11]R. Shi, H. J. Li, Z. Yang, and M. K. Kang, "Deposition mechanism of pyrolytic carbons at temperatures between 800-1200°C. Carbon, **35**(12), 1789-1792 (1997).

[12]F. Tuinstra and J. L. Koening, "Raman spectrum of graphite", J. Chem. Phys., **53**(3), 1126-1130 (1970).

[13]D. S. Knight and W. B. White, "Characterization of diamond films by Raman spectroscopy", J. Mater. Res., **4**(2), 385-393 (1989).

[14]C. A. Taylor, M. F. Wayne, and W. K. S. Chiu, "Microstructural characterisation of thin carbon films deposited from hydrocarbon mixtures", Surf. Coat. Tech., **182**, 131-137 (2004).

[15]O. Feron, F. L. Langlais and R. Naslain, "Analysis of the gas phase by in situ FTIR spectrometry and mass spectrometry during the CVD of pyrocarbon from propane", Chem. Vap. Deposition, **5**(1), 37-47 (1999).

[16]B. Descamps, G. L. Vignoles, O. Feron, J. Lavenac, and F. Langlais, "Kinetic modelling of gas-phase decomposition of propane: correlation with pyrocarbon deposition", J. Phs. IV, **11**, 101-108 (2001).

[17]W. Benzinger, A. Becker, and K. J. Hüttinger, "Chemistry and kinetics of chemical vapour deposition of pyrocarbon: I. Fundamentals of kinetics and chemical reaction engineering", Carbon, **34**(8), 957-966 (1996).

[18]J. Lavenac, F. Langlais, X. Bourrat, and R. Naslain, "Deposition process of laminar pyrocarbon from propane", J. Phys. IV, **11**, 1013-1021 (2001).

[19]Z. Jun and K. J. Hüttinger, "Mechanisms of carbon deposition-a kinetic approach", Carbon, **40**, 617-636 (2002).

[20]S. Eroglu and M. Gallois, "Chemical vapour deposition of pyrolytic carbon from impinging jets", Ceramics International, **22**, 483-487 (1996).

[21]A. Becker and K. J. Hüttinger, "Chemistry and kinetics of chemical vapour deposition of pyrocarbon - II. Pyrocarbon deposition from ethylene, acetylene and 1,3-butadiene in the low temperature regime", Carbon, **36**(3), 177-199 (1998).

[22]A. Becker and K. J. Hüttinger, "Chemistry and kinetics of chemical vapour deposition of pyrocarbon - III. Pyrocarbon deposition from propylene and benzene in the low temperature regime", Carbon, **36**(3), 201-211 (1998).

Ceramics for Advanced Nuclear and Alternative Energy Applications

STRENGTH TESTING OF MONOLITHIC AND DUPLEX SILICON CARBIDE CYLINDERS IN SUPPORT OF USE AS NUCLEAR FUEL CLADDING

Denwood F. Ross, Jr.
Gamma Engineering Corporation
15815 Crabbs Branch Way
Rockville, Maryland 20850

William R. Hendrich
Materials Science & Technology Division
Oak Ridge National Laboratory
Oak Ridge, TN 37831-6069

ABSTRACT

An important aspect of nuclear fuel rod cladding is its ability to withstand, throughout its lifetime, internal pressure loads during steady-state operation, transients, and accidents. Consideration is now being given to the use of silicon carbide duplex tubes as a replacement for the existing use of zirconium alloys. This project investigated the mechanical response of various silicon carbide tubular structures when subjected to internal pressurization using an expanding internal plug method. According to this test method, an incompressible polyurethane plug, fitted inside the test specimen, is subjected to axial compressive loading, which results in the internal pressurization of the test specimen.

Monolithic tubes manufactured were tested along with a series of fiber-reinforced composite tubular test specimens and duplex tubes comprised of a monolithic tube that is surrounded by a fiber-reinforced composite. The test results showed that:

- The monolith test specimens failed at about 0.2% tangential strain and the values of Young's modulus obtained were consistent with those published in the literature.
- The duplex test specimens also failed at about the same strain when the inner monolithic SiC failed. However, they exhibited a "graceful" mode of failure (no brittle fracture of the duplex test specimen) with the composite layer holding the structure in one piece.
- One set of composite-only test specimens exhibited a load-carrying capability lower than that of the monolithic SiC tubes, and tended to fracture in several pieces at low strain values.

INTRODUCTION

One important aspect of fuel rod cladding is its ability to withstand internal loads. Internal loads can arise from a pressure gradient between the inside of the cladding and the external environment. Internal pressure sources include the pre-fill gas pressure (Helium) and the accumulation of fission gases over the life of the fuel rod. Swelling of the fuel over long periods of operation may also provide a source of pressure loads on the cladding. External pressure, at power operation, is usually higher than the internal pressure. However, during certain transient and accident scenarios, there may be a positive gradient from the inside to the outside, which results in the tensile tangential loading of the cladding.

Another potential load arises from the reactivity-initiated accident, wherein the fuel pellet expands quickly due to the deposition of energy from the accident. The expansion could include both thermal expansion as a solid fuel pellet and expansion due to melting of some of the fuel. These various loads are postulated in the safety review of the fuel design [1], and thus it is necessary to know the cladding response.

Alloys of zirconium have been the cladding material of choice for most of the water-cooled reactors for almost half a century. These alloys have some common and desirable attributes, including:

- relatively inexpensive;
- generally compatible with operating environment;
- compatible with uranium dioxide
- strong;
- abundant.
- Relatively low neutron cross-section

However, as reactors proceed to ever-higher burnup levels, some problems with Zr have become apparent. The prolonged use in water reactor environment has produced surface corrosion to the extent that zirconium alloys interact less favorably during the postulated loss of coolant accident [2]. Further, test results are now showing that the consequences of the postulated reactivity insertion accident may be less favorable than previously thought, for these prolonged operating exposures. Although these negative factors may not absolutely preclude the use of Zr alloys for very high fuel burnup scenarios, there is nonetheless some interest in alternate materials for fuel cladding. For this reason, there are efforts to explore the use of various forms of Silicon Carbide (SiC) as a replacement for Zr tubing. SiC is an ideal candidate material for these applications because of its excellent mechanical, chemical and thermal properties, its ability to retain strength at elevated temperatures and most importantly its low neutron absorption cross-section.

One important aspect of programs to develop and qualify cladding materials is focused on determining their strength, which is the subject that concerns this report. The mechanical testing of the various SiC test specimens described in this report constitutes the first exploratory step in the direction of qualifying SiC for use as nuclear fuel cladding in a fission reactor. Several types of SiC tubes were fabricated for the purpose of testing the structural response to internal pressure. The test geometries include:

- Monolithic cylinders of Silicon Carbide
- Duplex test specimens, consisting of an inner monolith cylinder and a continuous SiC fiber-reinforced SiC matrix composite overlay.
- A continuous fiber-reinforced composite structure alone, with no inner monolithic cylinder

The general purpose of this test program was to determine the stress-strain relationships when the tubular structures are subjected to internal pressurization. Of interest was not only the stress and strain at failure up to the failure of the monolith (and the failure strain is quite small for the monolith) but also the behavior of the duplex test specimens after the initial failure of the interior monolith cylinder. The next phases of this study will include material properties at elevated temperatures, and the effects of radiation, and exposure to PWR water chemistry, on material properties. This paper will discuss some of the test results; a full description of the test specimens and test results is contained in reference 3.

TEST PROCEDURE

Tubular test specimens (monolithic, composite and duplex) were subjected to internal pressurization at ambient conditions according to the expanded plug test method [4]. The tests were carried-out using an electromechanical testing machine (MTS Model Alliance RT-50) and a test fixture consisting of a platen and two concentric pistons (Figure 1). The radial deformation of the test specimens was measured using a pair of capacitance proximity probes with a 0.8 mm range (Capacitec Mod HTP-75). Because these measurements require a conductive surface, it was necessary to apply a thin coating of chromium paint on the outer surface of the test specimens

Figure 1. Test set-up showing the test specimen, the loading pistons and the proximity gages

Plugs of incompressible polyurethane with a shore hardness of A95 were machined to a length of 12.5 mm and to a diameter slightly smaller than the inner diameter of the test specimen. The plug was mounted on the base pedestal and had a dowel interface to assure alignment with the piston. An initial small load was placed on the assembly to assure that the plug and specimen were centered and seated onto the base. The tests were performed under a constant crosshead displacement rate (0.017 mm/s). Tests on monolith specimens were terminated at the time of failure; however, for the duplex specimens, the crosshead was driven further until manually stopped (usually when the expanding specimen came close to the proximity gages). The test specimens were cut to a length of 63.5 mm using an instrumented saw equipped with a diamond blade.

Several types of SiC specimens were processed and tested as described below:

- Two monolithic tubes of SiC were tested. One was sintered α-SiC made by St. Gobain[1]. The second tube was CVD β-SiC made via CVD by TREX Enterprises[2].
- Three duplex tubes, identified as Round 4, were tested. These included test specimens made

[1] Niagara Falls, NY 14303
[2] Honolulu, HI 96813

by NovaTech[3] and Ceramic Composites Inc, (CCI)[4]. These tubes consisted of a 0.30-inch thick inner monolithic cylinder of sintered α-SiC and a composite outer layer consisting of continuous Hi-Nicalon fibers with a SiC matrix densified by chemical vapor infiltration. A helical winding architecture was used, with no crossover of tows that allowed for infiltration by CVI in between winding of the two layers. This was intended to reduce voids and delaminations found in previous rounds. A second purpose was to examine the CVI infiltration of tubes made with fiber tows with extra sizing, 7% by weight, as compared with the normal sizing provided on fiber tows of about 1% by weight. The idea was to increase the spacing between individual fibers in a tow during winding, by virtue of the thicker sizing, and then, after burnoff, this might allow less resistance to flow of the CVI gases into the interior of the tube. Nippon Carbon, the producer of CG-Nicalon fibers, produced a separate batch of fiber tows with high sizing specifically for this experiment. As a first step in the infiltration in the CVI reactor, methane gas was injected to provide a thin (less than 0.5 microns) pyrolytic carbon coating on the fibers. This interface layer was provided to assure composite behavior during mechanical loading. The inside diameter of Round 4 tubes was 0.330 inches; the wall thickness was 0.060 inches; and the outside diameter was 0.450 inches. The double composite layer was 0.030 inches thick.

- Three duplex tubes, identified as Round 5, made by NovaTech and CCI, were also tested. These tubes had a 0.030-inch thick inner monolithic cylinder of sintered α-SiC and a composite outer layer consisting of continuous Hi-Nicalon fibers with two different sizings, and a SiC matrix densified by chemical vapor infiltration. However, these tubes differed from the Round 4 tubes in that they used the "bamboo" fiber architecture, in which fibers are interwoven around the tube at 45° angles with tow crossover during winding. This winding architecture was chosen to increase the structural integrity, and resistance to delamination, of the composite layer. There was no separate CVI infiltration between layers, as there was with Round 4 tubes. However, the carbon interface layer was provided prior to CVI as in the round 4 tubes. These Round 5 duplex tubes had an inside diameter 0.355 inches; total wall thickness 0.040 inches, and outside diameter 0.435 inches. The composite layer was 0.010 inches.
- Starfire Systems[5] and NovaTech produced ten all-composite tubes for Gamma under funding provided by Westinghouse Electric Co. The steps involved in processing these tubes were as follows:

    - Coat 14-inch long graphite mandrels with release agent.
    - Dry wind the mandrels with bamboo pattern using Tyranno-SA fiber tows
    - Apply PyC fiber coating and SiC fiber polymer coating
    - Infiltrate matrix with Starfire polymer, cure and pyrolize at 1200°C for 8 hours. Repeat PIP cycles until no further weight is gained (about 8 cycles)

The finished Starfire tubes measured 0.371 inch ID, 0.420 inch OD, with a tube thickness of about 0.025 inches.

[3] Lynchburg, Virginia 24501
[4] Millersville, MD 21108
[5] Saratoga Technology + Energy Park. 10 Hermes Road. Suite 100, Malta, NY 12020

RESULTS FOR MONOLITHIC CYLINDERS

The data obtained during the mechanical tests consists of measurements of radial displacement of the outermost surface of the cylindrical test specimens, the axial compressive force recorded by the load cell and the piston displacement. From the radial displacement a tangential "strain" value was calculated according to:

$$\varepsilon_\theta = \frac{\Delta r}{r}$$

For data reduction and comparison with conventional material properties, the force on the piston[6] was converted to pressure, and the pressure then converted to stress, using the thin-cylinder approximation, i.e.-

$$\sigma_\theta = \frac{P}{t} r$$

where P is the internal pressure, r is the inner radius of the test specimen and t the wall thickness. For the linear portion of the initial deformation, values of Young's modulus were calculated. In general, the values obtained were consistent with those reported in the literature.

Results for two different types of monolith test specimens are shown in Figure 2. The test specimens failed in a brittle fashion, producing many small fragments during failure. For the TREX test specimen, the peak strain was 0.202% and the peak stress was 80 ksi. For the St Gobain test specimen, the peak strain was 0.224 % and the failure stress was 37.8 ksi.

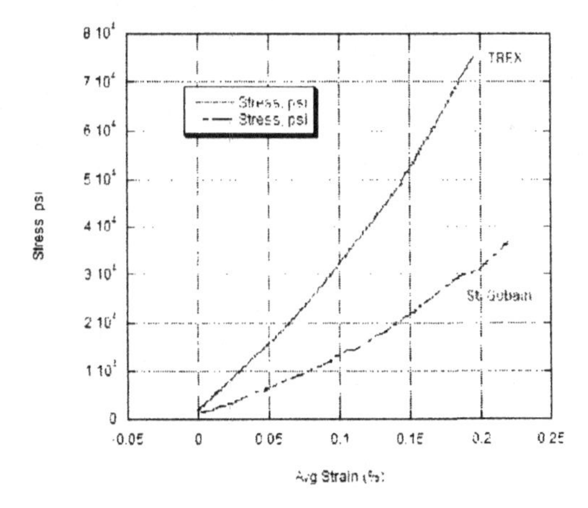

Figure 2. Stress-strain curves for two monolith cylinders

[6] The small load carried by the plug itself was generally negligible in reference to the failure load, and was not accounted for in the calculations.

RESULTS FOR DUPLEX CYLINDERS

The conversion from applied load (or pressure) to tangential stress for the duplex test specimens requires an assumption about the load-carrying properties of the duplex geometry. One assumption is that the inner and outer layers (that is, the monolith and the composite) react as a tightly coupled pair of annular cylinders, and follow an elastic deformation curve at the onset of loading. This somewhat arbitrary assumption is convenient for purposes of comparing one test specimen with another. However, it is tantamount to assuming that the modulus of the monolith (i.e., CVD material) is the same as the composite, and this is not the case. Work is underway to quantify the load sharing between monolith and composite during the initial loading (that is, before failure of the monolith).

Under this assumption, the results for duplex test specimens 4BP-1 and 5P-2 are shown in Figure 3. The peak stress that was calculated for both 4BP-1 and 5P-2 was 23.6 ksi. The first-failure strain was 0.083% for 4BP-1, and about 0.17% for 5P-2. The term first-failure refers to the unique behavior of the duplex test specimens that was seen when the inner monolith failed. During one time step, at the failure point, the load being carried dropped by about 60%. This corresponds to the cracking of the monolith. From that point on, the composite structure carried the entire load. The composite layer exhibited tough behavior and a graceful mode of failure, which are desirable attributes of this material for fuel cladding applications.

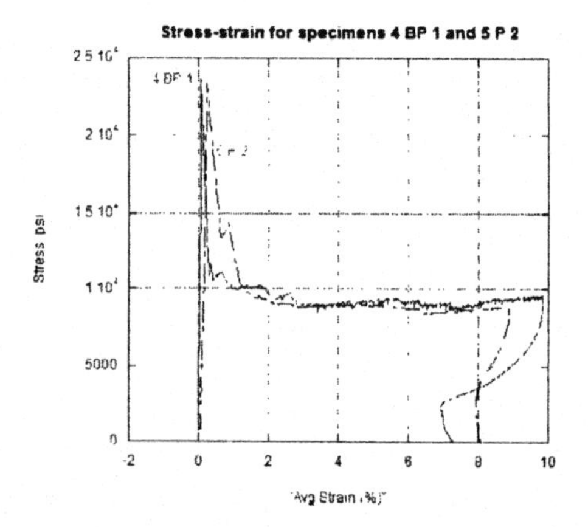

Figure 3. Stress-strain for test specimens 4 BP-1 and 5P-2

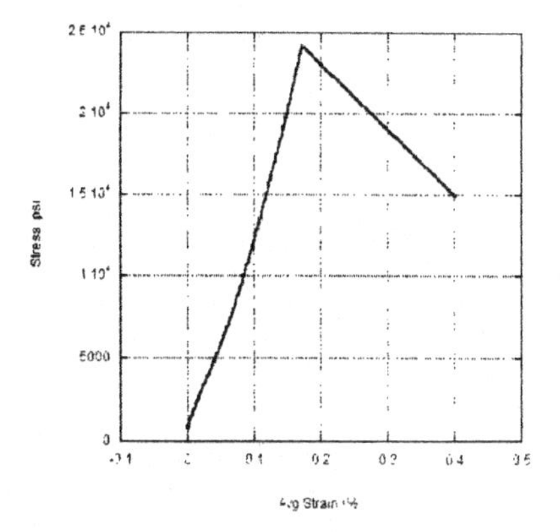

Figure 4. Stress-strain for test specimen 5P-1

The stress-strain relationship for specimen 5P-1 is illustrated in Figure 4 for the early part of the loading. The strain at first-failure was 0.172%. Although these three test specimens are considered identical, these differences in behavior illustrate the need to perform a large number of repetitions to obtain an appropriate statistical distribution of test results. This is particularly true for monolithic test specimens, which exhibit significant variability in strength

RESULTS FOR COMPOSITE-ONLY CYLINDERS

Starfire test specimen SN-010-1 failed in a brittle fashion, at a peak stress of only 6.9 ksi and at a strain of 0.5%. This failure mode was related to the inability of the fibers to debond and pullout. This behavior is illustrated in the SEM micrograph shown in Figure 6. The test specimen failed in a brittle manner and there is no evidence of fiber pullout, which would be associated with fiber debonding and fiber sliding. Fiber debonding and fiber sliding are the two main (and highly desirable) mechanisms responsible for the tough behavior exhibited by fiber-reinforced ceramic matrix composites. These mechanisms can be activated by tailoring the interface between the fibers and the matrix so that cracks that propagate through the matrix would be deflected at these interfaces. When this happens, the fibers can bridge the wake of the crack, preventing catastrophic failure. Interfaces in ceramic matrix composites are usually tailored by applying a thin, complain, coating to the fibers (e.g.- graphite, boron nitride) but it is clear from the scanning electron micrographs that such a coating was absent in this composite.

CONCLUSIONS AND INSIGHTS

The interpretation of the test results required, as stated earlier, some assumptions on load coupling and response for the duplex test specimens. This was further explored by comparing the load-strain results for a monolith test specimen and a duplex test specimen, which had an identical monolith inside. The results are shown in Figure 7. Here it is seen that the two response functions are almost identical, up to the first failure. This would seem to indicate that out to the strain at first failure (that is, about 0.2%) the responses are essentially the same, and thus the composite was not carrying much load at this point. However, after the failure of the monolith, the composite structure carries the entire load.

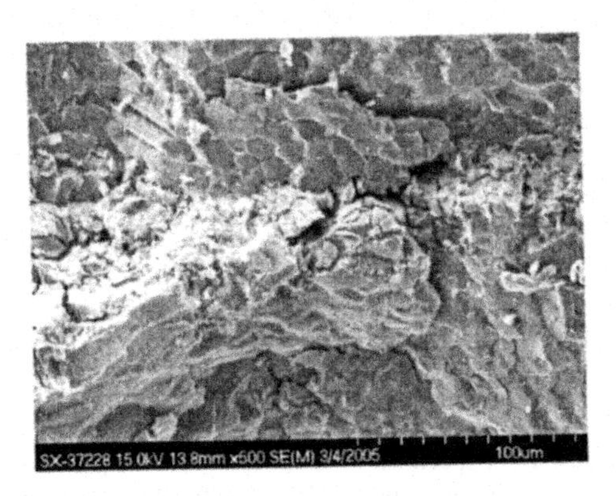

Figure 6. SEM of fracture surface of Starfire test specimen.

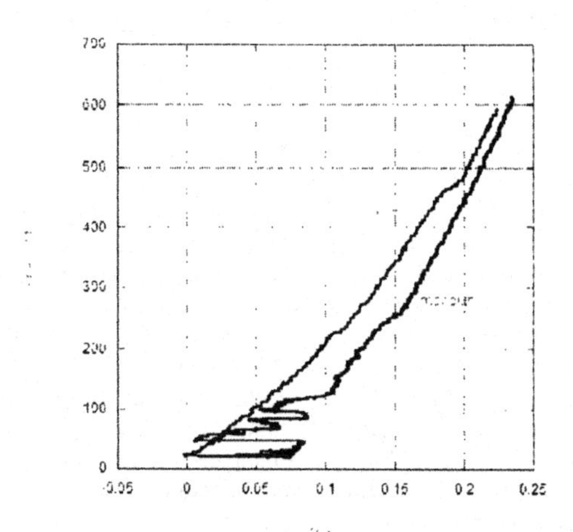

Figure 7. Comparison of load versus strain response of monolith and duplex geometry test specimens at low strains.

Some insights that have been gained from the strength testing thus far are:

- Monoliths tested alone will fail at low strains in a brittle fashion.
- Some types of duplex test specimens exhibit tough behavior and can experience large amounts of deformation and damage tolerance (graceful failure).
- The composite-only test specimens that were evaluated did exhibit poor performance because of lack of fiber debonding and sliding as a result of lack of an interfacial coating
- Although the number of tests have been limited, the data are reasonably consistent within a given test specimen geometry

Much more work needs to be done to make a more compelling case for using SiC structures as nuclear fuel cladding. Tests scheduled for calendar year 2006 include construction and initial utilization of an environmental test of SiC cladding materials in a combined radiation and reactor water chemistry environment. This is scheduled to be done at the reactor at MIT. Tests on the strength of SiC test specimens at elevated temperature are scheduled at ORNL.

ACKNOWLEDGMENTS

The work at Oak Ridge National Laboratory was sponsored by the Assistant Secretary for Energy Efficiency and Renewable Energy, Office of FreedomCAR and Vehicle Technology Program, as part of the High Temperature Materials Laboratory User Program, Oak Ridge National Laboratory, managed by UT-Battelle, LLC for the U.S. Department of Energy under contract number DE-AC05-00OR22725. The Silicon Carbide duplex test specimens provided by Gamma Engineering and Novatech were developed under DOE Small Business grants. The Starfire all-composite test specimens were produced by Starfire enterprises under contract to Gamma Engineering under a separate project sponsored by Westinghouse Electric Corporation. The authors would like to thank Herb Feinroth of Gamma Engineering and Edgar Lara-Curzio and Roger Jaramillo of Oak Ridge National Laboratory for technical guidance and for reviewing this manuscript.

REFERENCES

1. Nuclear Regulatory Commission Standard Review Plan 4.2: Fuel System Design Revision 3; April 1996; NUREG-0800
2. J-W. Yeon, Y. Jung, and S-I. Pyun, "Deposition behaviour of corrosion products on the Zircaloy heat transfer surface," Journal of Nuclear Materials 354 (2006) 163–170
3. H. Feinroth, et. al. "A Multi-Layered Ceramic Composite for Impermeable Fuel Cladding for Commercial Water Reactors." Gamma Engineering Report GN-54-03. November 2005.
4. W. J. McAfee, W. R. Hendrich and C. R. Luttrell, "Postirradiation Cladding Ductility Test Program," Oak Ridge National Laboratory Report ORNL/MD/LTR-228, March 200

SUBCRITICAL CRACK GROWTH IN HI-NICALON TYPE-S FIBER CVI-SIC/SIC COMPOSITES

Charles H. Henager, Jr.
Pacific Northwest National Laboratory
PO Box 999
Richland, WA 99352

ABSTRACT

SiC continuous-fiber composites are considered for nuclear applications but concern has centered on two major issues for this material at elevated temperatures in neutron environments. One is the differential materials response of the fiber, fiber/matrix interphase (fiber coating), and matrix under thermal-mechanical loads and irradiation-induced swelling. The other is subcritical crack extension when time-dependent and dose-dependent fiber creep can occur. In this study, both experiments and simulations were employed to understand and predict this behavior. Constant stress tests at elevated temperatures in inert environments without radiation damage are being used to explore subcritical crack growth in Type-S SiC-fiber composites. Additionally, a continuous-fiber composite is simulated by four concentric cylinders to explore the magnitude of radial stresses when irradiation swelling of the various components is incorporated. The outputs of this model were input into a time-dependent crack-bridging model to predict crack growth rates in an environment where thermal and irradiation creep of SiC-based fibers is considered. Under assumed Coulomb friction the fiber-matrix sliding stress decreases with increasing dose and then increases once the pyrocarbon swelling reaches "turn around." This causes an initial increase in crack growth rate and higher stresses in crack bridging fibers at higher doses. An assumed irradiation creep law for fine-grained SiC fibers is shown to dominate the radiation response, however.

INTRODUCTION

SiC-based continuous-fiber composites are considered for nuclear applications as structural components due to their high specific stiffness and strength, but concern has centered on the differential materials response of the fiber, pyrocarbon fiber/matrix interphase (fiber coating), and matrix under irradiation[1]. A modeling approach using concentric cylindrical regions to simulate continuous fiber composites to give the stress distribution in each region has been developed to explore thermo-mechanical loading effects[2-5]. This new model provided an elastic solution to the fiber, fiber-matrix interphase, matrix, and far-field composite regions for thermo-mechanical strains and irradiation-induced strains as functions of temperature and neutron dose[5]. The influence of pyrocarbon density on swelling was shown to determine the radial stress present at the fiber-matrix interface as a function of dose for composites containing stoichiometric SiC fibers, which can ultimately influence composite mechanical properties through the frictional sliding stress, τ.

A further concern for SiC composites at elevated temperatures is subcritical crack growth due to time-dependent fiber deformation[6-8]. Time-dependent, and thus dose-dependent, properties of interest for SiC-composites include retained strength, dimensional stability, and creep-crack growth[1], which have been address partly by a dynamic crack growth model developed to predict composite lifetimes due to assumed growth of internal cracks in these materials[6-8]. However, crack growth models have not been able to include these dose-dependent swelling data until the development of the modified four-cylinder model. The synthesis of these two models will allow a more detailed understanding and, hopefully, improved predictive capabilities with respect to time-dependent mechanical properties of SiC-composites.

In concert with models, crack growth experimentation with Type-S SiC fiber composites has been limited but it is this material that has shown the greatest stability under neutron irradiations up to 1273K and, thus, should be more carefully examined[1].

MODELING AND EXPERIMENT

Dose-Dependent Swelling Data and 4-Cylinder Model

An individual fiber with a fiber/matrix interphase coating embedded in an elastic matrix undergoing irradiation can be simulated by four concentric cylinders[3]. The surrounding composite is the outermost cylinder, while the matrix, fiber coating, and fiber are the remaining cylinders, with the fiber being the innermost cylinder. The cylinders are subject to three independent boundary conditions; axisymmetric temperature change, $\Delta T(r)$, uniaxial applied stress, σ_{oz}, and biaxial applied stress, σ_{or}, where r and z are the radial and axial components referred to cylindrical coordinates (r, θ, z). Stress relaxation was not allowed during irradiation or during cooling from the fabrication temperature and all components remain elastic and perfectly bonded. The fiber, matrix, and surrounding composite were treated as isotropic materials, while the pyrocarbon coatings were considered to be transversely isotropic[2]. The boundary conditions for the four-cylinder problem and solution are presented in[2,3] as the solution to eight simultaneous equations. The modification of the existing models is made by allowing all the regions to undergo a volume change due to irradiation-induced swelling[5].

With regard to the swelling of SiC matrix and SiC Type-S fiber the data of Price[9] for β-SiC swelling as a function of dose and temperature is used. For a typical fiber/matrix interphase of pyrocarbon, the data of Kaae[10] for the swelling of three pyrolytic carbons as function of fast-neutron fluence was used and converted to dpa by multiplying the fast-neutron fluence by 0.89. These curves are shown in Figure 1 for two types of pyrocarbon, denoted as high-density isotropic carbon (HDIC) and low-density isotropic carbon (LDIC)[10]. Using the material constants listed in Table 1, the 4-cylinder model predictions for the radial stress as a function of dose and pyrocarbon type are shown in Figure 2(a) for an assumed 100 nm-thick pyrocarbon layer. The radial stress that develops as a function of dose is a new parameter at the individual fiber level that can influence composite mechanical properties through changes in the frictional sliding stress. Assuming that the material obeys Coulomb's friction law, the frictional sliding stress can be related to the radial stress as

$$\tau = -\mu\sigma_{rr} \tag{1}$$

where τ is the frictional sliding stress, μ is the coefficient of friction at the fiber-matrix interface due to the pyrocarbon interphase, and σ_{rr} is the radial stress. This equation connects the output of the 4-cylinder model with fiber bridging mechanics, and the sliding stress as a function of dose is shown in Figure 2(b) for an assumed coefficient of friction of $\mu = 0.3$.

Dynamic Crack Growth Model and Fiber Creep

Pacific Northwest National Laboratory (PNNL) was among the first to identify and study time-dependent bridging in ceramic composites[11-13], which has resulted in a dynamic crack growth model and a crack growth mechanism map based on available experimental data as a function of temperature and oxygen partial pressure for continuous fiber composites with carbon interphases[8]. Once a relationship between crack-opening displacement and bridging tractions from crack-bridging elements is determined, a governing integral equation is obtained that relates the total crack opening, and the bridging tractions, to the applied load. The solution of this equation gives the force on the crack-bridges and the crack-opening displacement everywhere along the crack face[7]. This relation is rendered time-dependent by including appropriate bridging fiber creep laws and interface removal kinetics, if oxidation is an issue. For nuclear environments where neutron damage is occurring both thermal and irradiation-induced fiber creep are included but oxidation is not considered here. Since the frictional sliding stress, τ, is an input parameter for this dynamic model the results given above allow τ to be dose-dependent. The bridging model can be used to determine the effects of fiber creep and pyrocarbon type on composite mechanical properties in radiation environments.

Fiber thermal creep parameters for Hi-Nicalon Type-S SiC fibers were obtained from DiCarlo et al.[*], while radiation-induced creep parameters were obtained from the work of Scholz and Youngblood[14] for Sylramic SiC fibers, which is a fine-grained SiC fiber comparable to the Type-S fiber. These creep parameters are shown in Table 2. For comparison the creep parameters for Nicalon-CG and Hi-Nicalon fibers[15] are included. Each fiber follows a power-law creep equation with an activation energy, stress exponent, and time-temperature exponent that is characteristic of transient creep. The creep parameters exhibited by the various fibers are typical of either grain boundary sliding or interface-controlled creep.

Subcritical Crack Growth Experiments

SiC/SiC composites manufactured with Type-S fiber were obtained from GE Power Systems Composites having an 5-harness satin weave structure with 8-plies[†], a 150-nm thick pyrocarbon fiber coating, and an isothermal CVI SiC matrix. The resulting composite plate was 2.2 mm in thickness and approximately 230 x 150 mm in size with a density of 2.69 g/cm^3. Composite bars were cut from this plate having dimensions 50 x 4 mm and tested in –point bending with a fully articulated SiC bend fixture having 40 x 20 mm upper and lower loading point spacings, respectively. Fracture tests were also performed at two different strain rates, either fast at 1×10^{-4} s^{-1} or slow at 6.7×10^{-6} s^{-1}, in order to determine the envelope of temperature and strain rate where subcritical crack growth will occur in these Type-S composites. The subcritical crack growth tests were then performed on identical bend bars as the fracture tests except that the bars were loaded to a constant stress and held there for up to 3×10^6 s, or about 83 h and the deformation as a function of time was captured.

RESULTS AND DISCUSSION

Subcritical Crack Growth and Fiber Creep

The strain rate versus test temperature data taken from the 4-point bend strength testing of the Type-S SiC/SiC composites, shown in Figure 3, indicate that the onset of subcritical crack growth occurs at about 1273K, which is consistent with the knowledge of fiber creep of typical fine-grained SiC fibers. The fundamental working hypothesis of crack growth in continuous-fiber composites at elevated temperature in environments that are not aggressive towards either the fiber or the fiber/matrix interphase is that time-dependent extension of crack-bridging fibers

---

[*] JA DiCarlo, NASA Glenn Research Center, Cleveland, OH, personal communication.
[†] Type-S fabric was 18 epi, 500 filaments/yarn; lot number SCS0102, Nippon Carbon.

is controlling this process.[6-8] Creep of fine-grained SiC fibers, and other fibers, has been studied and documented by DiCarlo et al. and the following generic creep equation has been found to adequately represent the strain-time response of these fibers at elevated temperatures[15]

$$\varepsilon_c = A\sigma^n t^p \exp\left(\frac{-Qp}{RT}\right) \tag{2}$$

where ε_c is the creep strain, A is a constant, σ is the applied stress, t is elapsed time, T is temperature, Q is activation energy for creep, n is the stress exponent, p is the time-temperature exponent, and R is the gas constant. Values for these constants for typical SiC fibers are given in Table 2. Figure 4 shows strain-time curves for these SiC fibers and illustrates the improvements that have occurred in this field in terms of overall creep resistance. Figure 4 also illustrates that measurable thermal creep of Type-S fibers occurs at 1273K indicating that composites fabricated from this fiber should exhibit measurable subcritical crack growth at this temperature and above.

The subcritical crack growth data, shown in Figure 5, consists of 4-point bend strain data as a function of time at constant applied stress at the indicated temperature. These curves also appear to have a form similar to that of the single fibers and the composite deflection data can be fit to the same equation 2 as for the fibers. The best-fit parameters for n, p, and Q for the Type-S composite in inert environments are listed in Table 2. Data taken at 1573K was excluded as this specimen failed soon after loading and progressed too rapidly through the stable crack growth regime in order to contribute useful data.

To further support the hypothesis that crack growth is involved in the bend bar deflection optical micrographs taken in cross-section of the tensile face of one of the bend bars after testing are shown in Figure 6. At this point the bar is still intact but is uniformly cracked along the tensile face and the cracks are a few millimeters in length. These cracks grow by fiber creep of the bridging cracks present along the crack faces and a higher magnification photo reveals these bridging fibers intersecting the plane of the crack in Figure 6c.

Crack Growth Modeling

The dynamic crack growth model can provide excellent estimates of crack velocities and extension rates for single bridged cracks in a variety of loading configurations. More work is necessary before the model can explain the 4-point bend bar deflection data presented here since this requires addressing multiple cracks plus extending previous model results to link predicted crack growth with 4-point bar deflection data. However, the bend bar deflection data provides an adequate framework for the discussion of the growth of bridged cracks in Type-S composites for advanced nuclear energy applications.

The dynamic crack growth model reveals that Hi-Nicalon Type-S composites undergo slow crack growth at 1273K and above based on single fiber creep rate data. Bridged crack growth in Type-S composites are compared with identical cracks in Hi-Nicalon composites in Figure 7 and illustrate the improvements in fiber creep resistance that also influence crack growth rates as expected. The dynamic crack growth model predictions are in general agreement with experimental observations.

Using the assumed radiation-induced fiber creep law results in high crack growth rates, where the radiation-induced fiber creep is the dominant deformation mechanism. These crack growth results are shown in Figure 8 for a Type-S composite bar in 4-point bending at 1273K to 1473K with an assumed crack in the bar loaded to 10 MPa√m and with an assumed sliding stress value of 20 MPa. The data of Scholz is low dose data but it is anticipated that higher dose data in the future will reduce the irradiation creep contribution by perhaps a factor of 10. Irradiation creep should be weakly dependent on temperature and linear with dose rate and applied stress.

Variable Sliding Stress

The advantages of the 4-cylinder model are apparent when addressing the issue of pyrocarbon swelling and its effects on crack growth as in equation 1. The sliding stress is varied according to Figure 2(b) for both HDIC and LDIC pyrocarbon interfaces of 100-nm thickness based on knowledge of pyrocarbon swelling. For both pyrocarbon materials there exists a dose at which "turn around" occurs in their swelling and growth curves and the sign of the dimensional change reverses. Initially there is densification of the pyrocarbon and net radial shrinkage, which acts to reduce the sliding stress. At "turn around" pyrocarbon shrinkage stops and net radial swelling begins, which acts to put the fiber-matrix interface into compression and increase the sliding stress according to Equation 1. The sliding stress begins at 20 MPa at $t = 0$ and decreases with increasing tensile σ_{rr} but is not allowed to decrease below 5 MPa. These predicted crack growth curves are shown in Figure 9. Crack growth is initially more rapid compared to the constant sliding stress case, but then begins to decrease after "turn around" as the sliding stress begins to increase. Although the total crack length is longer for the variable sliding stress case, the crack growth rate has dropped below the rate for constant sliding stress after a dose of about 10 dpa.

Shown in Figure 10 is a plot of fiber bridging stress as a function of bridge position within the crack wake from the dynamic crack growth model. The nominal time-independent stress profile is modified due to fiber creep, which results in stress relaxation of fibers near the crack mouth relative to fibers at the crack tip. Incorporation of the variable sliding stress results in reduced bridging stresses compared to the constant sliding stress case and faster initial crack growth. With increasing dose, however, fiber-bridging stresses begin to increase rapidly compared to lower dose or compared to the constant sliding stress case. Although this acts to reduce the crack growth rate it may signify an impending problem with fiber failures with increasing dose.

CONCLUSIONS

The results of two different types of composite models that are linked with a common parameter, the fiber sliding stress (τ), are combined to the benefit of a dynamic crack growth model that was lacking detail for the dose-dependence of τ. Incorporating swelling data from two types of pyrocarbons, HDIC and LDIC, predicts that HDIC performs better compared to LDIC as a fiber-matrix interface material. Pyrocarbon swelling, however, is problematic and results in faster initial crack velocities for doses prior to "turn around" and slower growth rates after "turn around." The reduced crack velocity is encouraging but results in higher than expected fiber stresses as the sliding stress increases with increasing dose. This may argue against the use of pyrocarbon interfaces for fusion reactor environments if fiber stresses approach fiber fracture strengths. However, fast crack growth rates due to irradiation creep of SiC fine-grained fibers are of more concern. Additional fiber creep data at higher doses is needed to verify this concern. The experimental crack growth results indicate crack growth consistent with model predictions for Type-S fiber composites. The measured activation energies and bend bar time-dependent deflection data are similar to that for the creep of single SiC fibers.

REFERENCES

[1]Snead, L.L., et al., "Silicon carbide composites for fusion reactor application," *Advances in Science and Technology*, 33(10th International Ceramics Congress 2002, Part D), 129-140 (2003).

[2]Warwick, C.M. and T.W. Clyne, "Development of Composite Coaxial Cylinder Stress Analysis Model and Its Application to SiC Monofilament Systems," *J. Mater. Sci.*, 26, 3817-3827 (1991).

[3]Mikata, Y. and M. Taya, "Stress field in a coated continuous fiber composite subjected to thermo-mechanical loadings," *J. Composite Materials*, 19, 554-578 (1985).

[4]Kuntz, M., B. Meier, and G. Grathwohl, "Residual stresses in fiber-reinforced ceramics due to thermal expansion mismatch," *Journal of the American Ceramic Society*, 76, 2607-2612 (1993).

[5]Henager, C.H., E.A. Le, and R.H. Jones, "A model stress analysis of swelling in SiC/SiC composites as a function of fiber type and carbon interphase structure," *Journal of Nuclear Materials*, 329-333, 502-506 (2004).

[6]Henager, C.H., C.A. Lewinsohn, and R.H. Jones, "Subcritical crack growth in CVI SiCf/ SiC composites at elevated temperatures. Effect of fiber creep rate," *Acta Materialia*, 49, 3727-3738 (2001).

[7]Henager, C.H., Jr. and R.G. Hoagland, "Subcritical crack growth in CVI SiCf/SiC composites at elevated temperatures: Dynamic crack growth model," *Acta Materialia*, 49, 3739-3753 (2001).

[8]Jones, R.H. and C.H. Henager, Jr., "Subcritical crack growth processes in SiC/SiC ceramic matrix composites," *Journal of the European Ceramic Society*, 25, 1717-1722 (2005).

[9]Price, R.J., "Neutron irradiation-induced voids in β-silicon carbide," *J. Nucl. Mater.*, 48, 47-57 (1973).

[10]Kaae, J.L., "The mechanical behavior of Biso-coated fuel particles during irradiation. Part I: Analysis of stresses and strains generated in the coating of a Biso fuel particle during irradiation," *Nuclear Technology*, 35, 359-67 (1977).

[11]Henager, C.H., Jr., et al., "Time dependent, environmentally assisted crack growth in Nicalon-fiber-reinforced SiC composites at elevated temperatures," *Metall. Mater. Trans. A*, 27A, 839-49 (1996).

[12]Henager, C.H., Jr. and R.H. Jones, "Subcritical crack growth in CVI silicon carbide reinforced with Nicalon fibers: Experiment and model," *Journal of the American Ceramic Society*, 77, 2381-2394 (1994).

[13]Henager, C.H., Jr. and R.H. Jones, "High-temperature plasticity effects in bridged cracks and subcritical crack growth in ceramic composites," *Mater. Sci. Eng., A*, A166, 211-20 (1993).

[14]Scholz, R. and G.E. Youngblood, "Irradiation creep of advanced silicon carbide fibers," *Journal of Nuclear Materials*, 283-287, 372-375 (2000).

[15]DiCarlo, J.A., et al. Models for the thermostructural properties of SiC fibers. in High-Temperature Ceramic-Matrix Composites II. 1995.

Table 1: Properties used in calculations (typical values)

Materials	Young's Modulus (GPa)		Poisson's ratio		CTE ($10^{-6}C^{-1}$)	
	Axial E_L	Transverse E_T	Axial ν_L	In-plane ν_T	Axial α_L	Transverse α_L
Nicalon-CG fiber	200	200	0.2	0.2	8.2	8.2
Hi-Nicalon fiber	270	270	0.2	0.2	8.2	8.2
Type-S SiC fiber	420	420	0.2	0.2	4.0	4.0
SiC matrix	460	460	0.22	0.22	4.5	4.5
HDIC*	80	80	0.23	0.23	5	5
LDIC**	80	80	0.23	0.23	5	5

* High-density isotropic carbon
** Low-density isotropic carbon

Table 2: Fiber creep parameters

Fiber type and condition	Creep law[‡]	A (MPa^{-1} s^{-1})	K (MPa^{-1} dpa^{-1})	n	p	Q (kJ/mol)
Nicalon-CG Thermal	$\varepsilon = A\sigma^n t^p Exp\left(\frac{-Qp}{RT}\right)$	2		1.2	0.4	500
Hi-Nicalon Thermal	$\varepsilon = A\sigma^n t^p Exp\left(\frac{-Qp}{RT}\right)$	121		1.8	0.58	600
Type-S Thermal	$\varepsilon = A\sigma^n t^p Exp\left(\frac{-Qp}{RT}\right)$	0.0667		1.4	0.61	425
Sylramic Irradiation	$\varepsilon = K\dot{\phi}\sigma t$		4.7 x 10^{-6}	–	–	50
Bend Bar Strain-time Fit	$\varepsilon = A\sigma^n t^p Exp\left(\frac{-Qp}{RT}\right)$	2090		–	0.33	513

[‡] σ is stress in MPa, t is time in seconds, Q is activation energy in kJ/mol, A is a constant, and K is irradiation creep compliance for Sylramic fibers. Constants n and p are stress and time-temperature exponents, respectively.

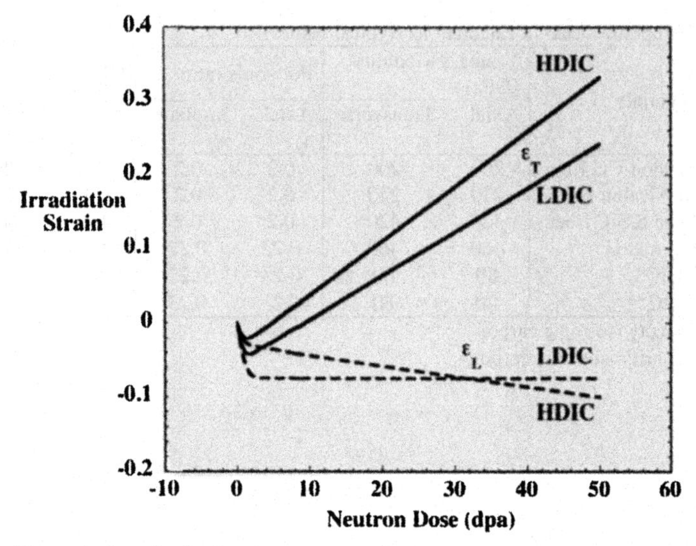

Figure 1. Longitudinal and transverse radiation-induced strain as a function of neutron dose in dpa for two types of pyrolytic carbon used as typical fiber coatings (HDIC and LDIC).

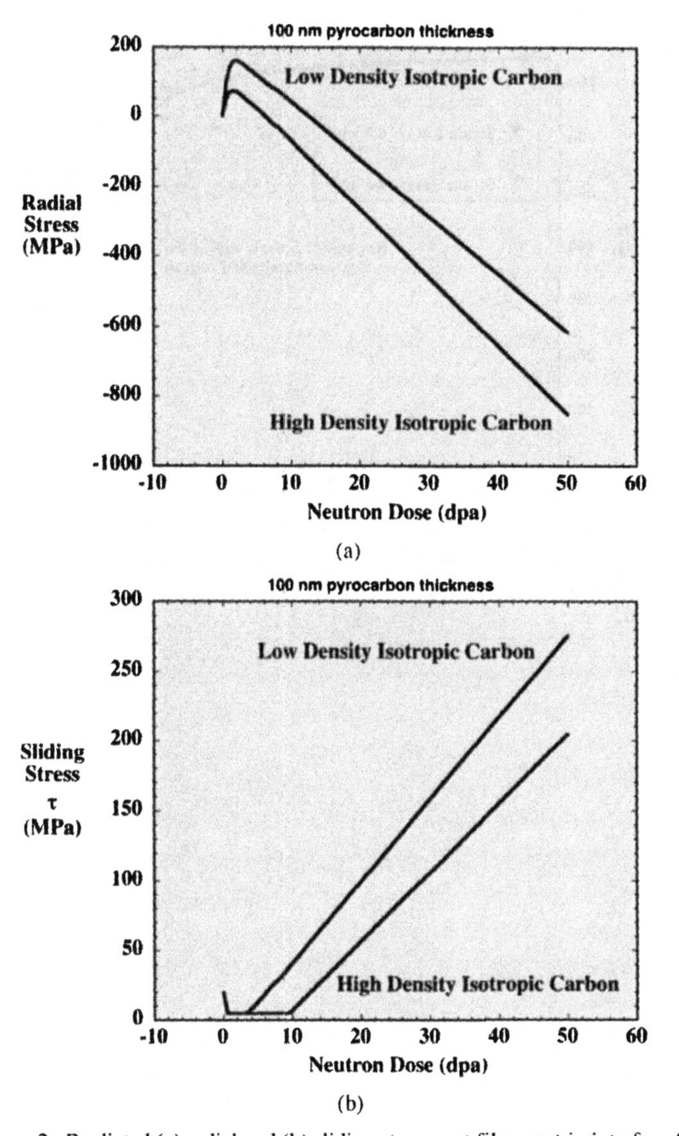

Figure 2. Predicted (a) radial and (b) sliding stresses at fiber-matrix interface for Type-S fiber composites with the indicated carbon coating of 100 nm subjected to neutron irradiation at 1273K.

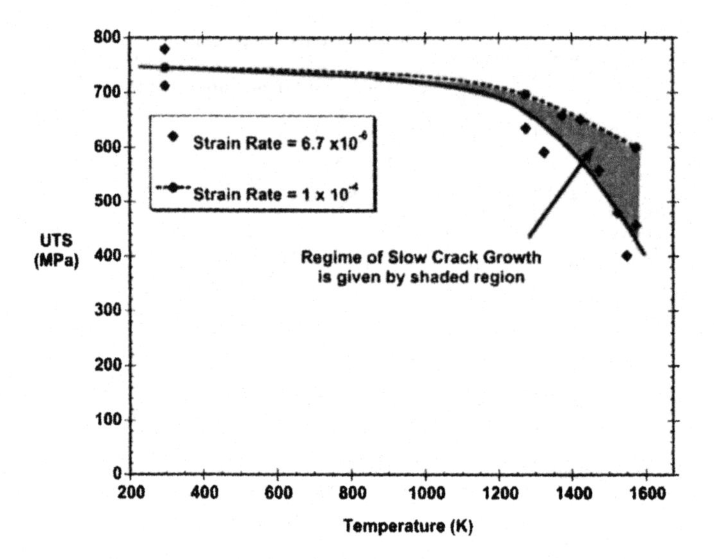

Figure 3. Fracture strength of Type-S SiC/SiC composite bars as a function of temperature and strain rate indicating temperature range where subcritical crack growth can occur.

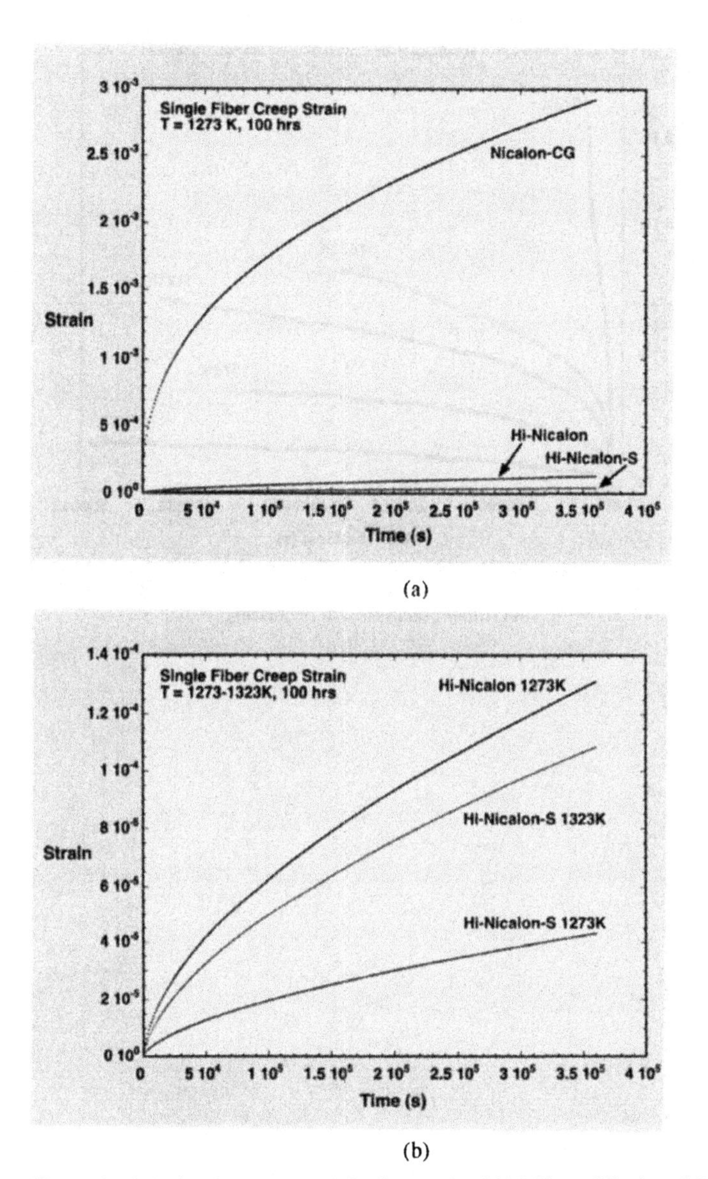

Figure 4. Calculated creep curves for fine-grained SiC fibers: Nicalon-CG, Hi-Nicalon, and Hi-Nicalon Type-S. In (a) all fibers are compared at 1273K and in (b) Hi-Nicalon and Type-S are compared. The improved thermal creep resistance of newer fibers is illustrated.

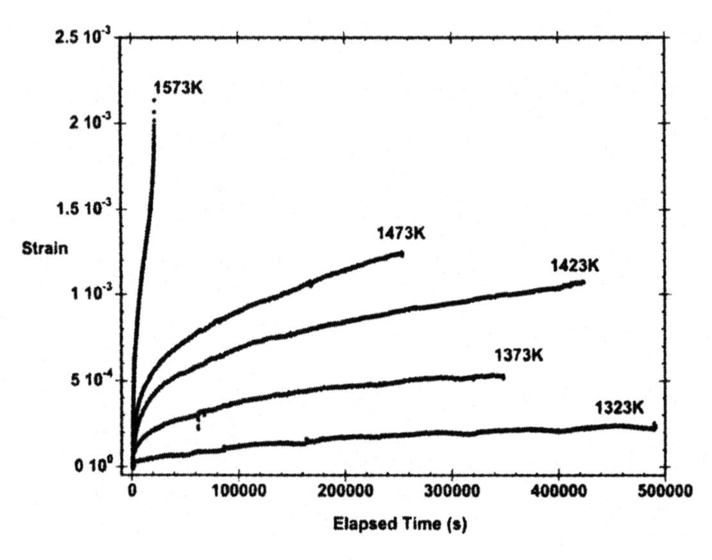

Figure 5. Constant stress bend strain data for Type-S SiC/SiC composite bars as a function of time showing subcritical crack growth occurring.

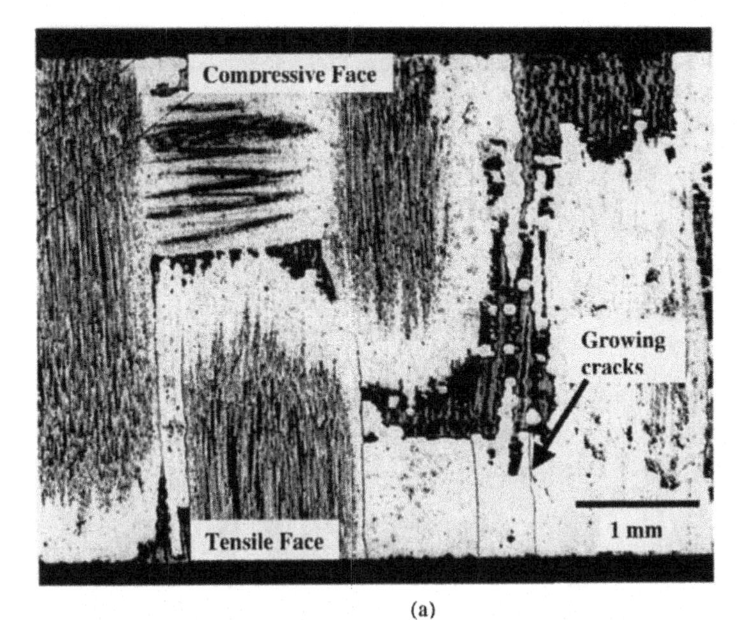
Compressive Face
Growing cracks
1 mm
Tensile Face

(a)

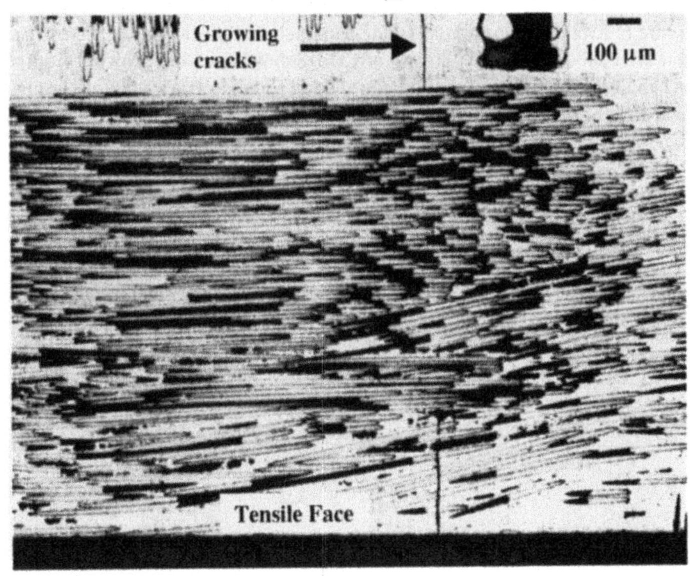
Growing cracks
100 μm
Tensile Face

(b)

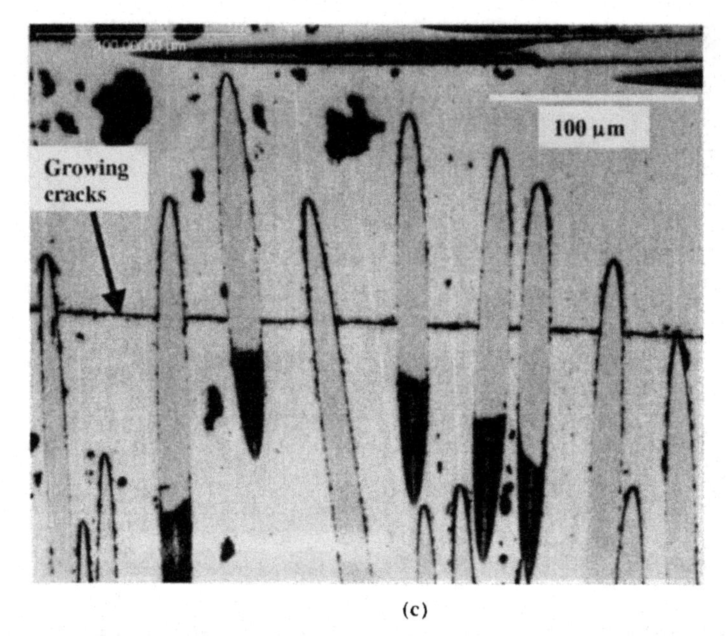

(c)

Figure 6. Optical micrographs (scale bars included) of (a) polished cross-section of bend bar after subcritical crack growth at 1373K showing cracks growing from tensile face of bend bar, (b) higher magnification section showing crack running through fiber bundle at 1473K, and (c) high magnification section showing crack bridging fibers at 1473K.

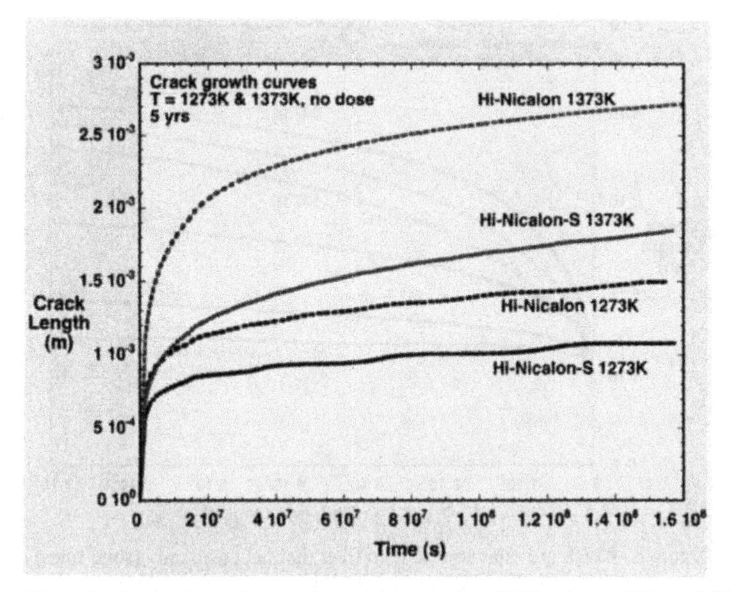

Figure 7. Dynamic crack growth model comparing Hi-Nicalon and Type-S fiber composites after simulated crack growth for 5 years under inert thermal conditions only.

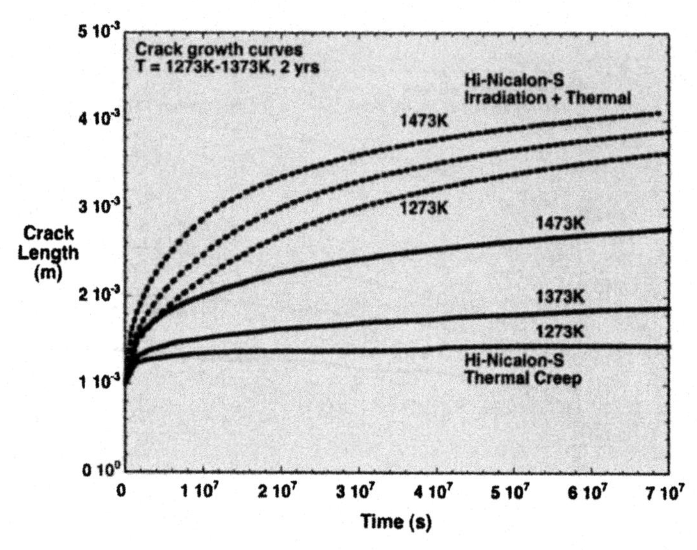

Figure 8. Crack growth curves due to fiber thermal creep only (solid lines) or thermal plus radiation-induced creep (dashed lines) at 1273K to 1473K.

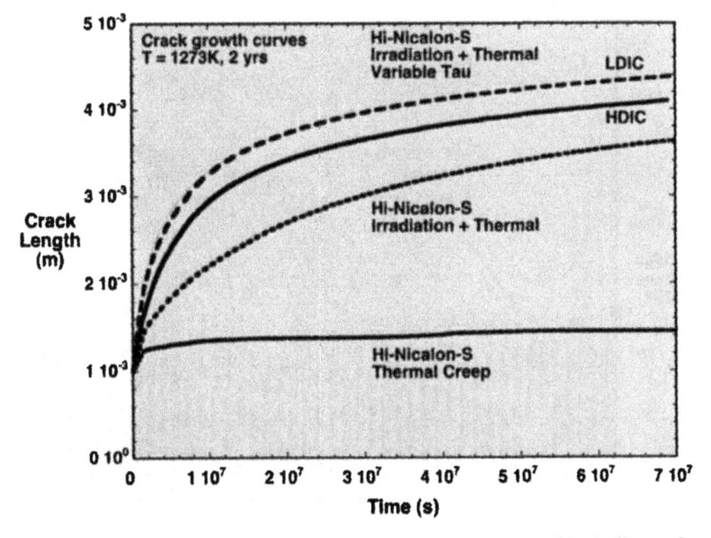

Figure 9. Crack length as a function of time showing the effects of including a dose-dependent sliding stress in the dynamic crack modeling at 1273K. Total time in seconds is for 2 years or 20 dpa.

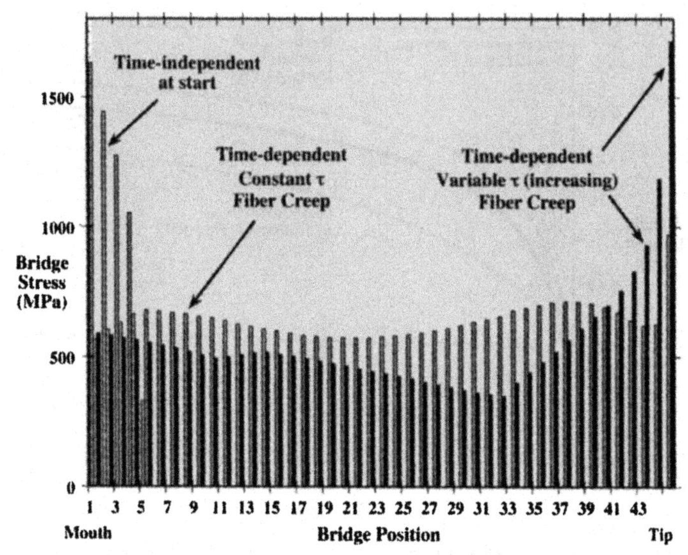

Figure 10. Plot of bridging fiber stresses as a function of position in crack-wake bridging zone and under the assumptions of constant or variable sliding stress. The time-dependent data is after 2 years of crack growth.

ELECTRICAL CONDUCTIVITY OF PROTON CONDUCTIVE CERAMICS UNDER REACTOR IRRADIATION

Tatsuo Shikama, Bun Tsuchiya, Shinji Nagata, and Kentaro Toh
Institute for Materials Research, Tohoku University,
2-1-1Katahira, Aobaku, Sendai, 980-8577 Japan
Tel; +81-22-215-2060, Fax; +81-22-215-2061, shikama@imr.tohoku.ac.jp

ABSTRACT

Possibility of converting radiation-energy such as energy of gamma-rays directly into electricity was surveyed with proton conductive ceramics in a fission reactor. Proton Conductive Ceramics, $SrCe_{0.95}Yb_{0.05}O_{3-\delta}$, $BaCe_{0.9}Y_{0.1}O_{3-\delta}$, and $CaZr_{0.9}In_{0.1}O_{3-\delta}$ were irradiated in a fission reactor, JMTR and their electrical conductivity and radiation induced electrical motive force were measured under irradiation. On some of specimens, a hydrated zirconium electrode was attached for studying effects of hydrogen on the electrical conductivity and the radiation induced electromotive force. An increase of the electrical conductivity was clearly observed at higher temperatures in specimens with a hydrated electrode. The experimental results are indicating that the electron conductivity is dominant below 400K and the conventional radiation induced electrical conductivity, whose magnitude is nearly proportional to the electronic excitation dose rate of radiation field, was observed. In the meantime, the base electron conductivity will be reduced by the radiation effects. Above 450-480K, the protonic conductivity is dominant in the present proton conductive ceramics and enhancement of proton diffusion was taking place under the irradiation. Increase of the electromotive force by the existence of hydrogen was not clearly observed in the present study, however, there was some possibility of contribution of the proton conduction to the RIEMF above 450K.

I. INTRODUCTION

Direct conversion of the nuclear radiation energy into the electricity has many advantages over the conventional thermal process. Several processes, such as collection of energetic fission fragments to electrodes and an electric power generation through a p-n junction of a semiconductor, have been studied but in vain, mainly due to severe degradation of electrical insulating ceramics separating collecting electrodes and microstructure-instability in a semiconductor by severe radiation damages.

Generation of an electric voltage of 0-5V and an electrical current up to about 1μA is usually observed under about a 1kGy/s ionizing (electronic excitation) dose rate, between two metallic parts separated by an electrical insulator, such as between a sheath tube and a core wire of a mineral insulating cable.[1] The phenomenon is called radiation induced electromotive force (RIEMF) and has been studied for a few tens years. A main cause of the RIEMF is thought to be a different influx and an outflow of electrons on two concerned metallic parts. As an influx and an outflow of electrons are in principle proportional to the ionizing dose rate, a generated current of the RIEMF is proportional to the electronic excitation dose rate with a high internal impedance, which is corresponding to the electrical conductivity of the insulator which is separating two metallic parts.

Thus, the electrical power generated by the RIEMF is in general marginal, being about 1μW (1μAx 1V) for a component of about $1cm^2$ and 1g, responding to a 1kGy/s ionizing dose rate, namely 1W/g. It can be effective to utilize dumping abundant radiation energy, though an energy conversion rate is only 10^{-6}. Several attempts can be proposed to enhance the RIEMF, such as utilization of abundant but low energy activated electrons, which usually do not contribute to the RIEMF. Utilization of radiation enhanced proton conductivity will be another method. In the present study, behaviors of electrical conductivity of proton conductive ceramics were studied under irradiation in a fission reactor, where the electronic excitation dose rate is high enough to detect its effect on the proton conduction. Behaviors of electrical conductivity are complicated in a fission reactor, and an electron irradiation and a 14MeV fusion neutron irradiation were also carried out to study fundamental aspects of radiation induced (enhanced) proton conductivity.

II. EXPERIMENTAL PROCEDURES

Three kinds of perovskite-type oxides, ytterbium doped strontium-cerium oxide ($SrCeO_3$), ($SrCe_{0.95}Yb_{0.05}O_{3-\delta}$), yttrium doped barium-cerium oxide ($BaCeO_3$), ($BaCe_{0.9}Y_{0.1}O_{3-\delta}$), and indium doped calcium-zirconium oxide ($CaZrO_3$), ($CaZr_{0.9}In_{0.1}O_{3-\delta}$), were irradiated in the Japan Materials Testing Reactor (JMTR) in the Oarai Research Establishment of Japan Atomic Energy Research Institute (JAERI; present Japan Atomic Energy Agency (JAEA)). Their electrical conductivity was measured in-situ under the reactor irradiation. The fast (E>1.0 MeV) and the thermal (E<0.683 eV) neutron fluxes were in the range of $1x10^{16}$-$6x10^{17}$ and $1x10^{17}$-$1.6x10^{18}$ n/m^2s, respectively. The associated gamma-ray dose rates were in the range 0.1k-2.0k Gy/s. Subcapsules, each of which was containing one specimen, were accommodated in an instrumented rig whose atmosphere was partially circulating purified helium of 2 atmospheric-pressure. The temperature was controlled by electric heaters as well as by changing a helium gas pressure between the subcapsule and the irradiation rig, after the reactor reached its steady-state operation mode at 50MW.

A measuring setup for the electrical conductivity was reported elsewhere.[2,3] The guard-ring configuration was adopted but the guard ring was connected to the ground potential through a wall of the subcapsule accommodating the specimen. Thin zirconium films were deposited on both sides of a plate-like specimen as electrodes and two platinum wires were connected to the two zirconium electrodes working as a cathode and an anode. The guard ring electrode was connected to a stainless-made subcapsule wall through a thick copper made specimen holder which is circumventing a leakage electrical current from coming into a current-measuring-side electrode. To study the protonic conductivity, 10keV proton ions were implanted into the anode side zirconium film and oxygen ions into the cathode side.

A similar irradiation was carried out in the Fast Neutronic Source (FNS) of the Tokai Research Establishment in the JAEA to study effects of 14MeV neutrons from the deuterium/tritium nuclear reaction, in an ambient atmosphere with the neutron fluxes in the range of $1x10^{9}$-$3x10^{12}$ n/m^2s, where an accompanying gamma-ray dose rate is very small.[4] 1.8-2.0 MeV electron-irradiation was also carried out in vacuum, in the Center of Investigation for Energy and Environmental Technologies (EURATOM/CIEMAT), measuring the electrical conductivity, the optical absorption and luminescence in-situ during the irradiation.[5]

III. Experimental Results and Discussions

(a) Electrical conductivity without hydrogen injection into zirconium electrode

A base electrical conductivity of the present perovskite-type oxides is relatively high and a radiation induced (enhanced) electrical conductivity was marginal when the hydrogen (proton) did not participate in the electrical conduction. Figure 1 shows the electrical conductivity of $SrCe_{0.95}Yb_{0.05}O_{3-\delta}$, increased by $8x10^{-9}$S/m from the base electrical conductivity under the 1.8 MeV electron irradiation[5] with a 10Gy/s dose rate, which is within the range of scatters of accumulated data of the radiation induced electrical conductivity (RIC) of several ceramic insulators shown in Fig. 2[6]. The base electrical conductivity decreased from $4.6x10^{-8}$S/m to about $5x10^{-10}$S/m after the 2.85MGy irradiation. The similar decrease of the base electrical conductivity was observed in the 14MeV neutron irradiation at 293 K and 373 K. The observed decrease of the base electrical conductivity is thought to be related with a radiation induced reduction, namely change of a charge state of ions in the oxides, in the case of $SrCe_{0.95}Yb_{0.05}O_{3-\delta}$, Ce^{+4} being reduced to Ce^{+3}.[5]

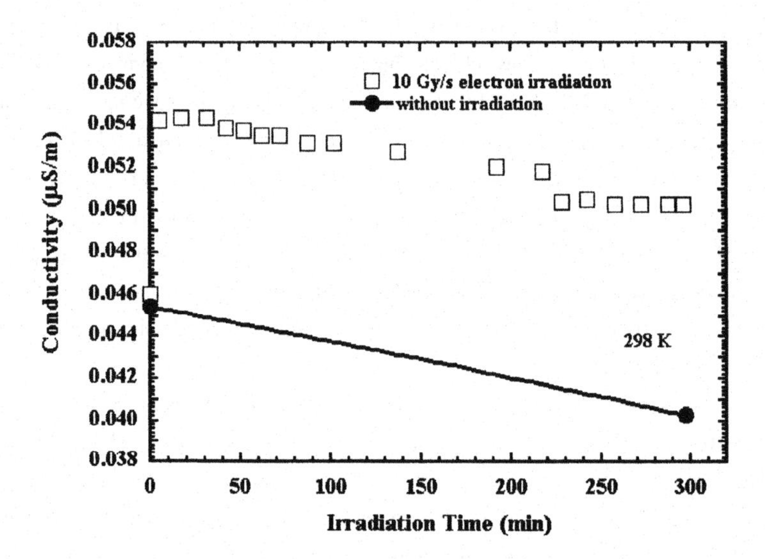

Figure 1 Electrical conductivity of $SrCe_{0.95}Yb_{0.05}O_{3-\delta}$ under 1.8 MeV electron irradiation at 297 K.

The reduction of the base electrical conductivity of $SrCe_{0.95}Yb_{0.05}O_{3-\delta}$ by the 1.8MeV electron irradiation is shown in Fig. 3 as a function of measuring temperature of the conductivity.[5] Temperature dependence of the electrical conductivity is divided into two components, one in a lower temperature range below 450K and the other above 450K. The reduction of the electrical conductivity is clearly responding to the reduction of the electrical conductivity of the lower temperature component. The lower temperature component is interpreted due to the electron (hole) conductivity[7] and the higher temperature components due to the proton conductivity. Thus, the electron irradiation suppressed the electron conductivity which is dominant below 450. As an electrolyte for a fuel cell, lower electron conductivity is essentially preferable. The proton conductivity is dominant down to room temperature in the irradiated $SrCe_{0.95}Yb_{0.05}O_{3-\delta}$ at 673K, which is indicating that the irradiated $SrCe_{0.95}Yb_{0.05}O_{3-\delta}$, can be used as an electrolyte in a fuel cell even at room temperature.

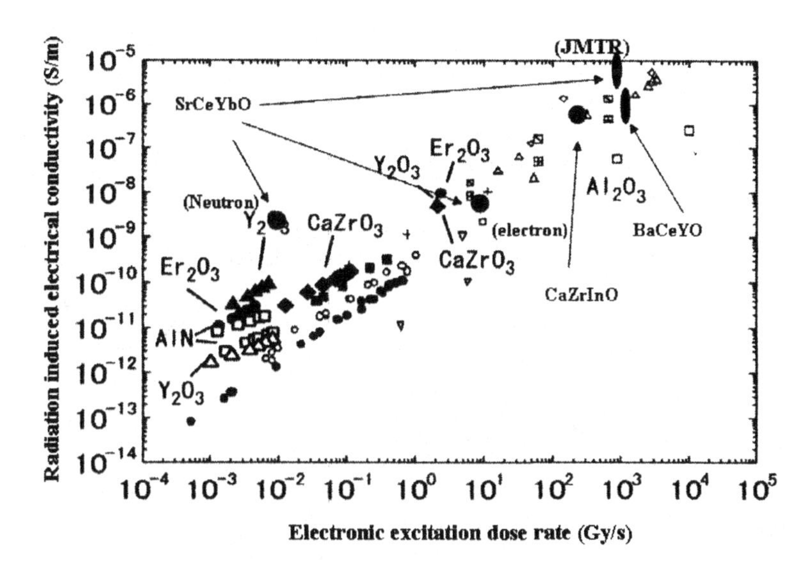

Figure 2 Electrical conductivity under the irradiation (RIC data compiled by T.Tanaka [6])

With decrease of the base electrical conductivity from $2x10^{-8}$S/m to about $(2-5)x10^{-10}$S/m, the radiation induced (enhanced) electrical conductivity of about $2x10^{-9}$S/m surfaced even in the 14MeV neutron irradiation at 293 K[4], after the irradiation up to about $2x10^{18}n/m^2$. The electronic excitation dose rate was estimated to be in the range of 10^{-2}Gy/s and the observed increase of the electrical conductivity was higher than the accumulated RIC data as shown in Fig. 2. Hydrogen atoms picked up from the atmospheric environment may play a role in the observed radiation induced (enhanced) electrical conductivity. The temperature dependence of the electrical conductivity under the 14MeV neutron is also plotted in Fig. 3. The temperature dependence is corresponding to that for the proton conductivity. Thus, the proton conduction may be observed even at room temperature under the 14MeV neutron. Effects of hydrogen will be discussed in the subsection (b).

At 1.9k Gy/s irradiation at 473K under the JMTR irradiation, the observed electrical conductivity of $SrCe_{0.95}Yb_{0.05}O_{3-\delta}$ was about $1x10^{-5}$S/m at the beginning of

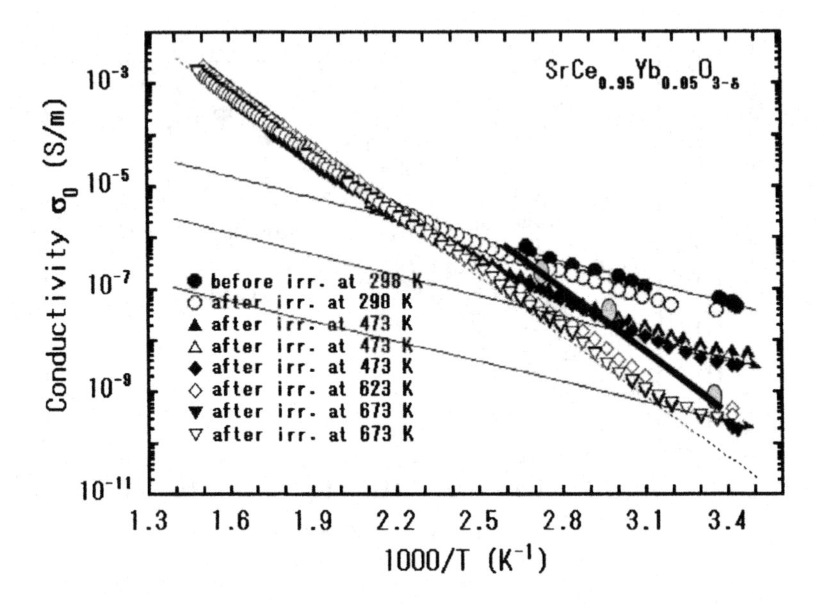

Figure 3 Electrical conductive ity of $SrCe_{0.95}Yb_{0.05}O_{3-\delta}$ after 1.8MeV electron irradiation, as a function of measuring temperature. The electrical conductivity measured in-situ during 14MeV electron is also shown as .

irradiation and $1x10^{-6}$S/m at the $2.8x10^{23}n/m^2$ fast neutron fluence. The electrical conductivity of $1.2x10^{-6}$ at the beginning of the irradiation and $6x10^{-7}$S/m at $3.6x10^{23}n/m^2$ fast neutron fluence, at 2k Gy/s were observed for $BaCe_{0.9}Y_{0.1}O_{3-\delta}$ at about 520K, and $8x10^{-7}$S/m at 230 Gy/s for $CaZr_{0.9}In_{0.1}O_{3-\delta}$ at about 420K, all of which are also within the scatter of accumulated RIC data[6] as shown in Fig. 2. The electrical conductivity measured below 450K on the specimens without hydrogen implantation depended nearly proportionally on the reactor power, being shown in Fig. 4, as will be described in the next section. This also agreed with the conventional RIC, as will be described below. Decrease of the electrical conductivity in the course of irradiation was also observed in the JMTR irradiation in general as described above.

(b) Effect of hydrogen implantation into anode zirconium film and temperature dependence of electrical conductivity

The electrical conductivities of $BaCe_{0.9}Y_{0.1}O_{3-\delta}$ with hydrogen implanted in its anode zirconium film and without hydrogen implantation (hereafter denoted as a specimen with hydrogen and a specimen without hydrogen) are shown in Fig. 4 at the startup of JMTR as a function of the ionizing dose rate. It is clearly shown that the specimen with hydrogen had larger electrical conductivity than that without hydrogen. Other two oxides, $SrCe_{0.95}Yb_{0.05}O_{3-\delta}$ and $CaZr_{0.9}In_{0.1}O_{3-\delta}$ showed the similar results. The hydrogen implantation into an anode increased the electrical conductivity. The observed increase of electrical conductivity with increase of the ionizing dose rate in the specimens without hydrogen was nearly proportional to the increase of the reactor power (the increase of the neutron flux and the electronic excitation dose rate), as described in the previous section. This suggested that the normal RIC is dominant, because the temperature was below 450K. In the meantime, the increase of the electrical conductivity in specimens with hydrogen was not linearly dependent on the reactor power, suggesting that the proton conduction might contribute to the electrical conductivity at higher reactor powers.

Figure 5 summarizes temperature dependence of the electrical conductivity of $BaCe_{0.9}Y_{0.1}O_{3-\delta}$.[9] As discussed in the previous section for the temperature dependence of the electron irradiated $SrCe_{0.95}Yb_{0.05}O_{3-\delta}$, for non-irradiation conditions (a bench top experiment), the electrical conductivity could be composed of two mechanisms as shown for $SrCe_{0.95}Yb_{0.05}O_{3-\delta}$, one having a lower apparent activation energy of about 0.2 eV (mechanism 1; electron conduction) in the temperature range below 480K, and the other having a higher apparent activation energy of about 0.7-0.8 eV (mechanism 2; proton conduction).[8,10] The hydrogen implantation into the anode did not alter the electron conductivity substantially but increased the proton conductivity by about the two orders of magnitude. Under the JMTR irradiation, measurements of temperature dependence of the electrical conductivity below about 400 K could not be realized due to a large nuclear heating rate and the electrical conductivity by the mechanism 1 could not be evaluated during the JMTR steady-state operation. Without the hydrogen implantation into the anode, the increase of the proton conductivity by 2k Gy/s irradiation was about 7 times. In the meantime, with the hydrogen implanted anode, the observed increase of proton conductivity by the 1.1 kGy/s irradiation was more than two orders of magnitude. Under the JMTR irradiation, further increase of the electrical conductivity (mechanism 3) from that by the mechanism 2 was observed at higher temperatures, namely, above 623 K for the specimen without hydrogen and above 523 K for the specimen with hydrogen. This increase of the electrical conductivity at higher temperatures was not observed in the bench top experiments as can be seen in Fig. 5.

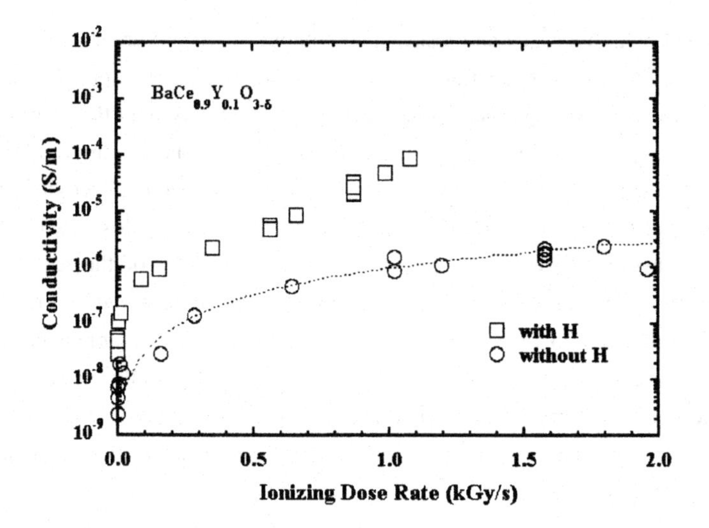

Figure 4 Electrical conductivity of $SrCe_{0.95}Yb_{0.05}O_{3-\delta}$ at JMTR startup

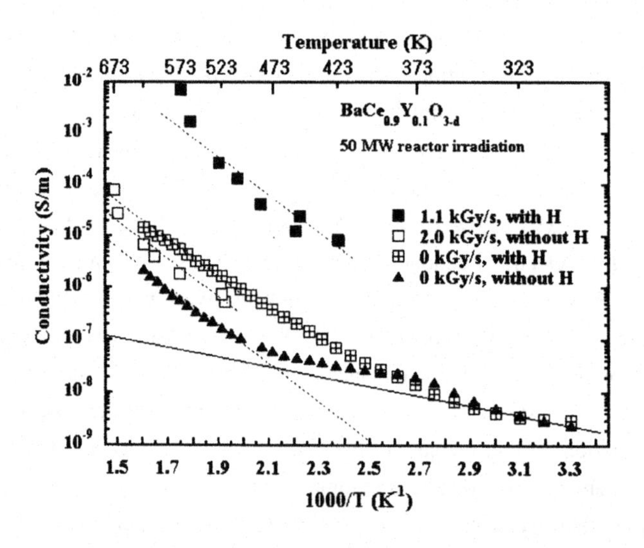

Figure 5 Temperature dependence of electrical conductivity of $BaCe_{0.9}Y_{0.1}O_{3-\delta}$, without irradiation and under JMTR irradiation.[9]

Without hydrogen, the electron conduction will be dominated below 450 K for irradiated $SrCe_{0.95}Yb_{0.05}O_{3-\delta}$, and below 480-500K for $BaCe_{0.9}Y_{0.1}O_{3-\delta}$, thus, the RIC, which have very weak temperature dependence, may be observable below 450 K. A qusi-linear dependence of the observed electrical conductivity on the electronic excitation dose rate below 450 K in the specimen without hydrogen shown in Fig. 4 will support this hypothesis.

Results showed that the proton conductivity was enhanced substantially under the irradiation. As the apparent activation energy was not affected by the irradiation, a concentration of a charge carrier, namely a proton, must be increased by the irradiation, assuming that the mechanism 2 is corresponding to the proton conduction. Thus, the irradiation will enhance ionization of hydrogen and penetration of protons into the oxide. Increase of the electrical conductivity by the mechanism 3 is strongly suggesting that mobility of ions composing the oxide is also increased by the irradiation. The appearance of the mechanism 3 at lower temperatures in the specimen with hydrogen will imply that the existence of protons will enhance the increase of ion mobility by the irradiation.

Enhancement of the ion migrations by the irradiation was also implied by the surface morphology change by the 14 MeV neutron irradiation even at 293 K. Small pits of a few μm in diameter on the surface of $SrCe_{0.95}Yb_{0.05}O_{3-\delta}$ were smoothened by the irradiation up to $1x10^{19}n/m^2$, suggesting a long-distance migration of constituent ions at 293 K. In the case of the electron irradiation at 293 K, formation of SrO fine needles on the $SrCe_{0.95}Yb_{0.05}O_{3-\delta}$ surface was observed suggesting decomposition of $SrCe_{0.95}Yb_{0.05}O_{3-\delta}$ into SrO and $Ce_2O_{4-\delta}$.[5]

(c) Radiation Induced Electromotive Force (RIEMF)

Radiation induced electromotive force (RIEMF) is an electrical current and voltage generated between two electrically conductive material separated by an electrically insulating material under irradiation. A voltage of 0.1-5V and a current of 0.1-several μA is a typical RIEMF observed in an about $100mm^2$ size sample under about 1kGy/s irradiation.[1] Figure 6 shows an electrical current (RIEMF) generated between two electrodes attached to the both sides of $CaZr_{0.9}In_{0.1}O_{3-\delta}$ under the JMTR irradiation. A small increase of the RIEMF current was observed by the hydrogen implantation to the anode. However, the observed increase of the RIEMF current was marginal. Also, the RIEMF in the components having a proton conductive insulator was comparable to generally observed one, where alumina or magnesia is used as an insulator. Thus, the present results showed that the proton conductive insulator will not enhance the RIEMF and the proton conductivity will not contribute to the RIEMF in general.

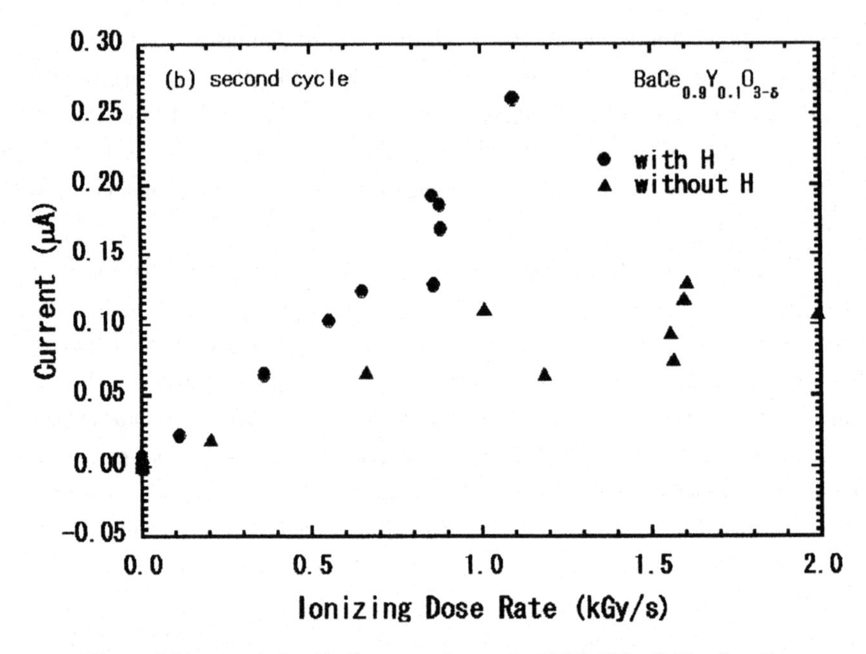

Figure 6 Observed electrical current due to the RIEMF in $CaZr_{0.9}In_{0.1}O_{3-\delta}$ under JMTR irradiation

However, it should be noted that the irradiation temperature was about 400K at the ionizing dose rate of 1kG/s, just below the temperature where the proton conduction became dominant without irradiation. Above 1kGy/s, where the temperature was above 400K, increase of the RIEMF in the specimen without hydrogen deviated from the linear relationship as shown in Fig. 6. There is some possibility that the RIEMF having a negative polarity existed above 400K. Figure 7 shows the RIEMF observed in $SrCe_{0.95}Yb_{0.05}O_{3-\delta}$ without hydrogen as a function of the reactor power, where the temperature also rose with increase of the reactor power. At 10MW, where the ionizing dose rate was about 1kG/s and the irradiation temperature exceeded 400K, magnitude of the RIEMF tuned to decrease and changed its polarity at 30MW. Among several possibilities, contribution of the proton conduction may be cause of the change of polarity of the RIEMF. Optimization of the configuration of two electrodes, a proton conductor and a holder surrounding them may realize larger RIEMF at elevated temperatures.

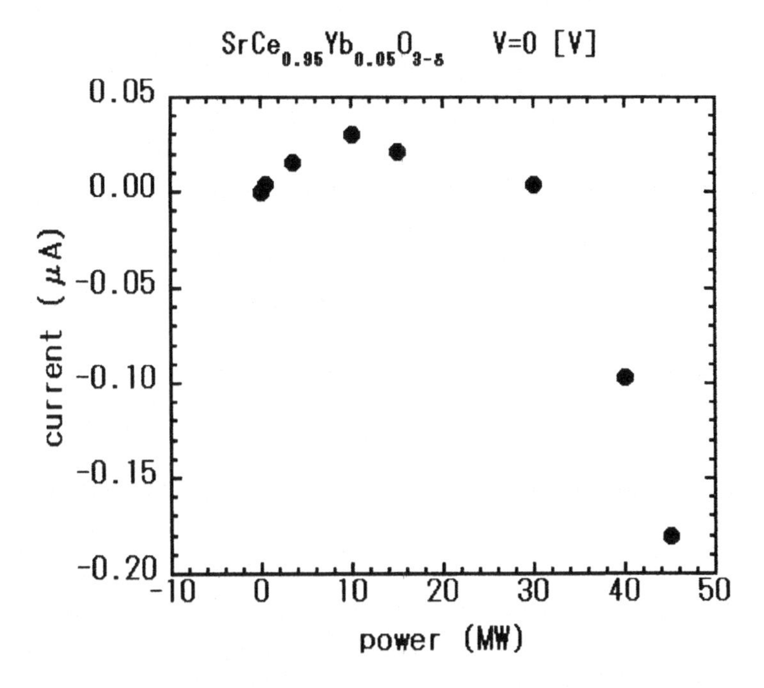

Figure 7 RIEMFof $SrCe_{0.95}Yb_{0.05}O_{3-\delta}$ without hydrogen as a function of JMTR reactor power. Temperature increased with increase of reactor power and exceeded 400K.

IV CONCLUSION

Behaviors of electrical conductivity of perovskite-oxides were reviewed, under 1.8-2.0 MeV electrons, 14 MeV neutrons and a fission reactor irradiations. Effects of hydrogen (protons) on the electrical conductivity were specially studied. All of the studied perovskite-oxides had relatively large base electrical conductivity and the normal RIC by the excited electrons was marginal in these oxides under the irradiation. The protonic conduction which was dominant above about 450 K was strongly enhanced by the irradiation. In the meantime, base electronic conduction was suppressed by the radiation induced reduction, in the case of $SrCe_{0.95}Yb_{0.05}O_{3-\delta}$, Ce^{+4} to Ce^{+3}. Existence of protons in the oxides and the radiation effects will enhance the long-distance migration of ions and phase changes will take place at relatively low temperatures in the present oxides. The present proton conductive insulators did not generate large RIEMF.

REFERENCES

[1]T. Shikama, M.Narui, and T.Sagawa, Nucl. Instr. Methds in Phys. Res., B122 (1997) 650-656

[2]B. Tsuchiya, S. Nagata, K. Saito K. Toh and T. Shikama, to be published (2006).

[3]T.Shikama, Advances in Science and Technology, 24 (1999) 463.

[4]B. Tsuchiya, S. Nagata, K. Toh, T. Shikama, M. Yamauchi and T. Nishitani, Fusion Science and Technology, 47, 4 (2005) 891.

[5]B. Tsuchiya, A. Moroño, E. R. Hodgson, T. Yamamura, S. Nagata, K. Toh and T. Shikama, Phys. Stat. Sol. (c)2, No. 1, (2005) 204.

[6]T.Tanaka, T.Shikama, M.Narui, B.Tsuchiya, A.Suzuki and T.Muroga, Fus. Eng. Des., 75-79 (2005) 933-937

[7]H. Iwahara, Solid State Ionics, 77 (1995) 289.

[8]H. Uchida, H. Yoshikawa, T. Esaka, S. Ohtsu and H. Iwahara, Solid State Ionics, 36 (1989) 89.

[9]B.Tsuchiya et al.,presented at the 12th Int. Conf. on Fusion Reactor Materials, Santa Barbara, California, December, 2005.

[10]T. Arai, A. Kunimatsu, K. Takahiro, S. Nagata, S. Yamaguchi, Y. Akiyama, N. Sata, and M. Ishigame, Solid State Ionics, 121 (1999) 263.

THE EFFECTS OF IRRADIATION-INDUCED SWELLING OF CONSTITUENTS ON MECHANICAL PROPERTIES OF ADVANCED SIC/SIC COMPOSITES

Kazumi Ozawa
Graduate School of Energy Science, Kyoto University
Gokasho, Uji, Kyoto 611-0011, Japan

Takashi Nozawa
Metals and Ceramics Division, Oak Ridge National Laboratory
Oak Ridge, TN 37831-6138

Tatsuya Hinoki, Akira Kohyama
Institute of Advanced Energy, Kyoto University
Gokasho, Uji, Kyoto 611-0011, Japan

ABSTRACT

The effects of irradiation-induced swelling of constituents on tensile properties of advanced SiC/SiC composites were examined. After the swelling of the fiber and matrix ion-irradiated up to 1dpa at 1000°C was measured, the modified four phase model was applied in order to estimate the residual stress in each constituent. The swelling mismatch was obtained between Hi-Nicalon™ Type-S fiber (0.27%) and ICVI-matrix (0.48%), which increases the residual stresses according to the model. These increases of the residual stresses were roughly consistent with the tensile stress-strain behavior of the same SiC/SiC composites after neutron irradiation up to 1dpa at 1000°C and the behavior of the misfit stress obtained by the hysteresis loop analysis methodology, which is related with the residual stress. These results suggest that irradiation-induced residual stress caused by the differential swelling of each constituent may affect the tensile properties in this irradiation condition.

INTRODUCTION

Silicon carbide (SiC) continuous fiber-reinforced SiC matrix composites (SiC/SiC composites) are attractive structural materials for fusion reactors and gas fast reactors because of their superior mechanical properties at high temperature, low susceptibility to radiation damage[1].

One of the most important issues of SiC/SiC composites under irradiation is the difference of swelling among each constituent, because it can cause the degradation of interfacial and mechanical properties. SiC/SiC composites reinforced with less crystalline and non-stoichiometric SiC fibers have shown the interfacial debonding due to mismatch of swelling behavior of the fiber (densification) and β-SiC matrix (swelling) after neutron irradiation, resulting in the degradation of mechanical properties[2, 3]. In contrast, due to similar swelling of β-SiC matrix and highly crystalline and near-stichiometric SiC fibers such as Hi-Nicalon™ Type-S[4] (Nippon Carbon Co., Ltd.) and Tyranno™-SA[5] (Ube Industries, Ltd.), advanced SiC/SiC composites have exhibited excellent irradiation resistance in ultimate bend/tensile strength[3, 6-8].

Nevertheless, it is still very important to understand the swelling behavior of each constituent of even advanced SiC/SiC composites in order for the precise prediction and evaluation of irradiation resistance. In our previous work[8], the tensile properties of advanced SiC/SiC composites with pyrolytic carbon (PyC) interphase after neutron irradiation up to 1dpa at 1000°C were examined. They also exhibited good irradiation resistance in tensile strength, but the change of misfit stress, which was related with residual stress, was obtained from the hysteresis loop analysis methodology. This change implies that there was a swelling mismatch between the advanced fiber and matrix; however, there are few data[6] about the swelling behavior of these materials. In addition, there have been detailed residual stress analysis of advanced SiC/SiC composites[9], but this was based on the assumption that the swelling of the advanced fiber and matrix is the same.

The main purpose of this study is to figure out the irradiation-induced swelling of SiC constituent of advanced SiC/SiC composites, and to clarify the effects of swelling mismatch on tensile property changes after irradiation, taking into account of thermal residual stresses induced by coefficients of thermal expansions (CTE).

EXPERIMENTAL PROCEDURE

Materials

The materials used were advanced SiC/SiC composites reinforced with unidirectional, near-stoichiometric SiC fibers; two types of fiber were prepared, Hi-Nicalon™ Type-S (hereafter HNLS) and Tyranno™-SA Grade-1 (hereafter TySA). They were fabricated by isothermal chemical vapor infiltration (ICVI) method at Hyper-Therm High-Temperature Composites, Inc. for the ORNL/Kyoto University round robin irradiation program[10]. The fiber/matrix interphase was PyC/SiC multilayer $(PyC_{20}/SiC_{100})_5$ in order to avoid the swelling of PyC during the ion-irradiation. For ion irradiation experiments, the samples were cut into 1mm×3mm×0.2mm pieces and the surface was polished by 1μm diamond powders. The irradiation surfaces were chosen to be parallel to fiber direction.

Ion irradiation

The ion irradiation of the samples was carried out at the DuET facility in Institute of Advanced Energy, Kyoto University[11]. In order to induce displacement damages, 5.1MeV Si^{2+} ions were irradiated (single-ion irradiation). The depth profiles of damage was calculated by TRIM-98 code[12], using a sublattice-average displacement energy of 35 eV and the density of SiC of 3.21 g/cm^3, as shown in Fig. 1. Displacement damage level, irradiation temperature, and irradiation dose rate were up to 1dpa, 1000°C, and 1×10^{-3}dpa/s, respectively. The temperature of the irradiated surface was measured with high-resolution thermography during irradiation.

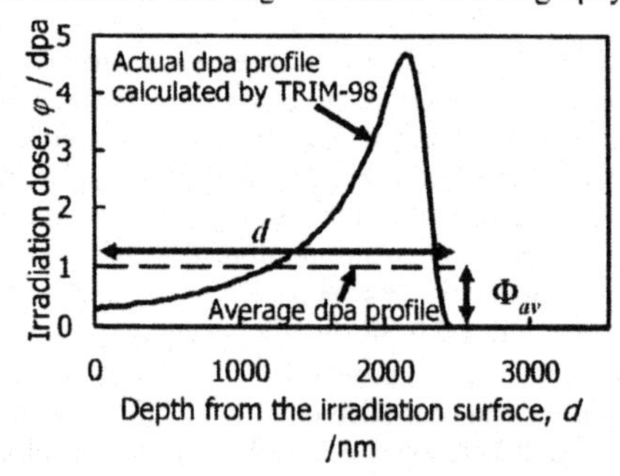

Fig. 1 The depth distribution of displacement damage level in SiC calculated by TRIM-98.

Post ion irradiation experiments

After the ion irradiation, the degree of swelling was determined by means of a precision surface profilometry utilizing meshes to facilitate measurements. The surface height change as a consequence of irradiation-induced volume expansion was measured with the Micromap™ interferometric optical surface profiling system (Micromap Inc., Tucson, USA). The method of ion-irradiation was as follows. First, the meshes which have landscape holes were set perpendicular to the fiber direction in order to avoid unnecessary restriction as shown in Fig. 2 (a). Second, the ion irradiation was performed (Fig. 2 (b)), and finally the step height was measured (Fig. 2 (c)). The step heights of fibers were measured at the center of isolated fibers which had nearly the original fiber diameter. The step heights of matrix were also measured at the area far from fibers and sample edges. The degree of swelling was evaluated as

$$\Delta V/V(\Phi_{av}) \cong \Delta Z/d \tag{1}$$

where Φ_{av} is the average irradiation dose, ΔZ is the step-height, and d is the length of irradiated region. The standard experimental procedure is described elsewhere[13].

For a microstructural investigation by cross-sectional transmission electron microscopy (TEM), the irradiated samples were subjected to a thin foil processing using a focused ion beam (FIB) device (Micrion/JEOL JFIB 2100). The thinning procedure using FIB is described in detail elsewhere[14]. The XTEM observation was performed using a JEOL JEM-2010 at 200kV.

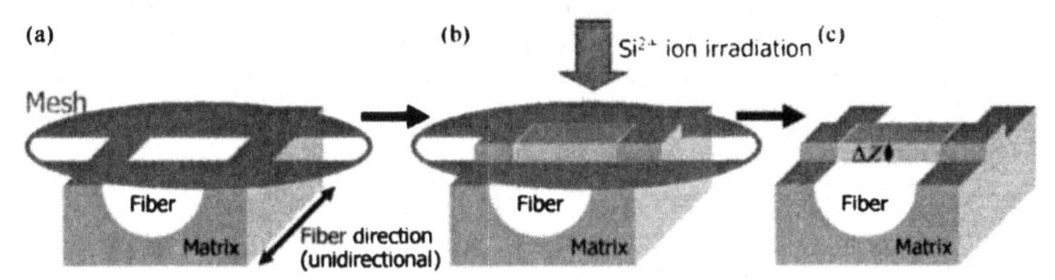

Fig. 2 Schematic diagrams of the method of ion-irradiation.

RESIDUAL STRESS ANALYSIS

Four phase model

In order to evaluate residual stress in each constituent of the advanced SiC/SiC composites, the modified four phase model[9] based on Refs.[15-17] was applied. According to Ref.[9], the residual stresses in radial (r), hoop (θ), and axial (z) directions in the components are expressed as

$$\sigma_r^{(n)} = C_{11}^{*(n)}\left(A_n - \frac{B_n}{r^2}\right) + C_{12}^{*(n)}\left(A_n + \frac{B_n}{r^2}\right) + C_{13}^{*(n)}E - \beta_1^{(n)}\Delta T - \gamma_1^{(n)} \tag{2a}$$

$$\sigma_\theta^{(n)} = C_{12}^{*(n)}\left(A_n - \frac{B_n}{r^2}\right) + C_{11}^{*(n)}\left(A_n + \frac{B_n}{r^2}\right) + C_{13}^{*(n)}E - \beta_1^{(n)}\Delta T - \gamma_1^{(n)} \tag{2b}$$

$$\sigma_z^{(n)} = 2C_{13}^{(n)}A_n + C_{33}^{(n)}E - \beta_3^{(n)}\Delta T - \gamma_3^{(n)} \tag{2c}$$

where n is the number of the respective phase (1: composite, 2: matrix, 3: interphase, 4: fiber), $C_{ij}^{(n)}$ are the material elastic properties, ΔT is the uniform temperature change from the fabrication temperature, and A_n, B_n, and E are the coefficients of the solutions of the four phase model problem for the appropriate boundary conditions. The detail of calculating these coefficients is given elsewhere[16]. $\beta_i^{(n)}$ are defined as

$$\begin{aligned} \beta_1^{(n)} &= (C_{11}^{(n)} + C_{12}^{(n)})\alpha_T^{(n)} + C_{13}^{(n)}\alpha_A^{(n)} \\ \beta_3^{(n)} &= 2C_{12}^{(n)}\alpha_T^{(n)} + C_{33}^{(n)}\alpha_A^{(n)} \end{aligned} \tag{3}$$

where $\alpha_T^{(n)}$ and $\alpha_L^{(n)}$ are the transverse (T) and longitudinal (L) coefficients of thermal expansion (CTE) for each component. $\gamma_i^{(n)}$ are the irradiation-induced stresses defined as

$$\begin{aligned} \gamma_1^{(n)} &= (C_{11}^{(n)} + C_{12}^{(n)})\varepsilon_T^{(n)} + C_{13}^{(n)}\varepsilon_A^{(n)} \\ \gamma_3^{(n)} &= 2C_{12}^{(n)}\varepsilon_T^{(n)} + C_{33}^{(n)}\varepsilon_A^{(n)} \end{aligned} \tag{4}$$

where $\varepsilon_T^{(n)}$ and $\varepsilon_L^{(n)}$ are determined from swelling data for each constituent in the composite. For the composite phase, the properties are calculated by the rule of mixture using the volume fractions of matrix, interphase, and fiber phases. These are given by rations of the rations of the various component radii as

$$V_m = 1 - \left(\frac{r_3}{r_2}\right)^2, \quad V_c = \left(\frac{r_3}{r_2}\right)^2 - \left(\frac{r_4}{r_2}\right)^2, \quad V_f = \left(\frac{r_4}{r_2}\right)^2 \tag{5}$$

where V_m, V_c and V_f are the volume fraction of matrix, interphase, and fiber, r_1, r_2, r_3 and r_4 are the radius of composite, matrix, interphase and fiber phase, respectively. The fiber, matrix, and composite phases were treated as isotropic materials, while the interphase phase was considered to be transversely isotropic.

Parameters used in the analysis

For this analysis, parameters of materials used were listed in Table I. The stresses were calculated at 1 and 7.7dpa. For swelling of SiC constituents, the values obtained from this ion-irradiation experiment were used. The swelling from 1dpa to 7.7dpa is assumed to be same, because the swelling of CVD-SiC was nearly saturated around 1dpa[18, 19]. For swelling / shrinkage of the PyC interphase, the data of Kaae and co-workers[20, 21] was used (high-density of isotropic carbon, HDIC[20]). A residual thermal stress was present due to the ΔT of -1100°C from a fabrication temperature of composites. Values of the phase radii were chosen to match microstructural information for advanced SiC/SiC composites; equivalent to composite containing either HNLS or TySA with a uniform fiber radius of 6 or 5 um, respectively and a fiber volume fraction of 0.29[10]. The thickness of PyC interphase was 720nm[8, 10].

Table I Lists of material parameters used in the analysis.

Materials	Young's modulus [GPa]	Poisson's ratio	Thermal expansion coefficient [$10^{-6}/^{o}C$]
Hi-Nicalon™ Type-S	420	0.2	5.1
Tyranno™-SA Grade-1	320	0.2	4.5
Isotropic PyC	20	0.23	5.0
ICVI-SiC	460	0.21	4.0

RESULTS AND DISCUSSION

Swelling of each SiC constituent after ion-irradiation

Table II exhibits the swelling after ion irradiation up to 1dpa at 1000°C. As shown in Table II, the differential swelling between HNLS and ICVI matrix was obtained (HNLS: 0.27%, ICVI-SiC: 0.48%), while such remarkable mismatch was not detected between TySA (0.46%) and ICVI matrix. In addition, the swelling of TySA in this experiment was nearly same as the data calculated from the density change after neutron irradiation up to 7.7dpa at 800°C, reported by Hinoki et *al.*[6].

Table II The swelling values of SiC constituent after ion-irradiation up to 1dpa at 1000°C. Numbers in parenthesis show standard deviations.

Materials	Swelling [%]	Number of measured
Hi-Nicalon™ Type-S	0.27 (0.102)	10
Tyranno™-SA Grade-1	0.46 (0.082)	10
ICVI-SiC	0.48 (0.096)	40
CVD-SiC[19]	0.64 (0.005)	-

Figure 3 exhibits the TEM micrographs of each SiC constituent after irradiation up to 1dpa at 1000°C. From the TEM microstructural observations, no irradiation-induced void (< 5nm in diameter) was detected in each specimen after irradiation, although small and sparse voids (< 5nm) were observed in CVD-SiC irradiated up to 10dpa at 1000°C[19]. Hence, it is hard to consider that void swelling affects these differences. It is speculated that these differences might arise from the difference of grain size (i.e. the minimal distance between a grain boundary and next one) as shown in Fig. 4, and/or pre-existing pores as sinks of point defects.

Residual stress analysis

Figure 5 exhibits the residual radial, hoop, and axial stresses in HNLS and TySA composite systems after irradiation up to 1 and 7.7dpa at 1000°C. The residual stresses in composites were strongly dependent on the both CTE and swelling mismatch in this model. As shown in Fig. 5 (a), the thermal residual stresses induced by CTE mismatch occurred in the unirradiated HNLS composite: 438MPa of axial tensile stress in fiber and 223MPa of axial compressive stress in matrix, and 156MPa of tensile radial stress at the PyC interphase. These residual stresses were increased by swelling mismatch of fiber, matrix and PyC interphase after irradiation up to 1dpa (Fig. 5 (b)). Especially, of importance is the increase of residual stress in axial and radial directions; this increase might affect the tensile properties, as discussed in the following section. On the contrary, residual stresses (except axial one in the PyC interphase) were reduced after irradiation up to 7.7dpa. This result of increase and decrease in residual stresses with increasing dose was consistent with the result reported by Henager Jr. et *al.*[9]. This may be due to the behavior of PyC in c-axis; HDIC[20, 21] in c-axis shrinks to a maximum shrinkage value (at ~2dpa-SiC). then turns around and begins to swell (at 5dpa-SiC~). About 800MPa of tensile axial stress in the PyC interphase occurred by irradiation up to 1 and 7.7dpa, possibly due to swelling mismatch of SiC constituents (swelling) and the PyC (shrinkage) in this direction.

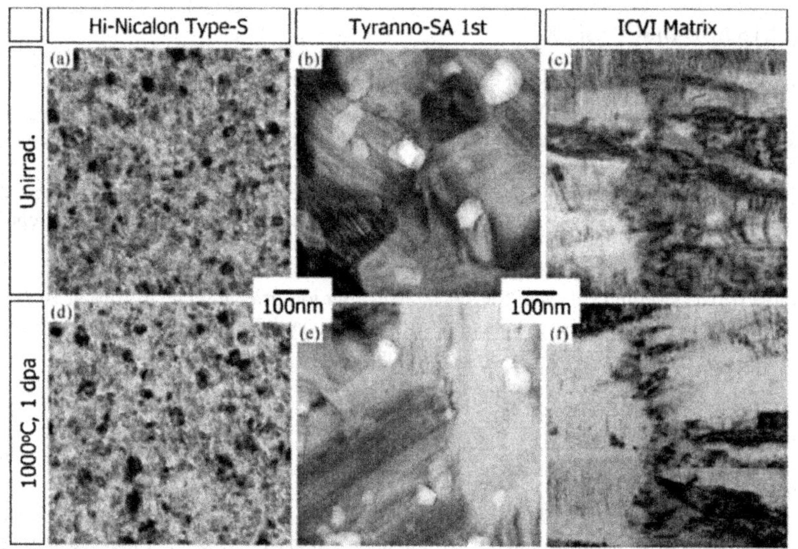

Fig. 3 TEM micrographs of each SiC constituent after ion-irradiation up to 1dpa at 1000°C.

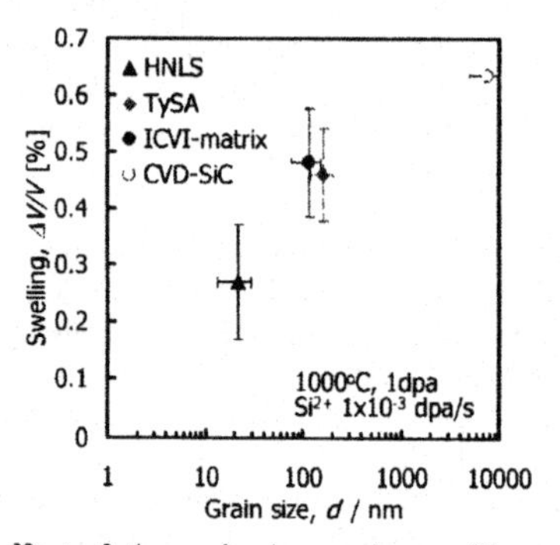

Fig. 4 Irradiation effect of the grain size on the swelling of SiC constituents.

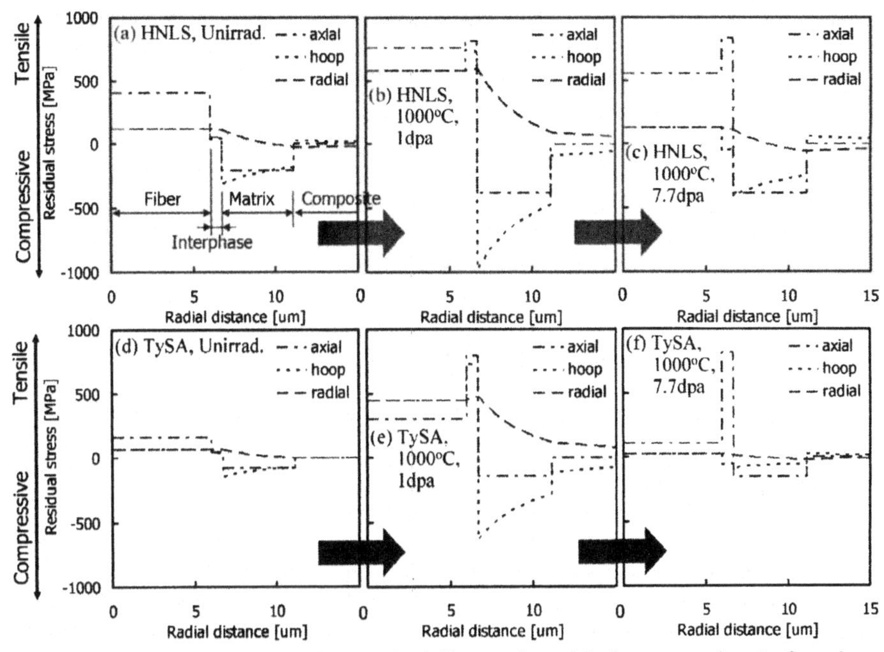

Fig. 5 Irradiation effects of the differential swelling on the residual stresses using the four phase model analysis. In this figure, HNLS, and TySA means the Hi-Nicalon Type-S/PyC/ICVI matrix, and Tyranno-SA Grade-1/PyC/ICVI matrix, respectively.

The residual stresses in the TySA composites after irradiation behaved as were the case with the HNLS composite. but the degree of change in these stresses was smaller than the HNLS composite due to smaller CTE and swelling mismatch.

Comparison with tensile test after neutron irradiation

In our previous work[8], the tensile properties of the same SiC/SiC composites (Hi-Nicalon^{TM} Type-S/PyC/ICVI) with PyC interphase after neutron irradiation up to 1dpa at 1000°C were examined. The important results of this previous work were as follows, depicted in Ref. [8]: The composites with even thick PyC interphase exhibited the good irradiation resistance; decrease of proportional limit stress (PLS) was small. and increase of ultimate tensile strength (UTS) was obtained. In addition, in order for the further investigations. the hysteresis loop analysis methodology[22-24] proposed by E. Vagaggini and co-workers was also carried out. In result, a slight degradation of interfacial sliding stress and large increase of misfit stress in negative was estimated.

These results were roughly consistent with the results of the residual stress model analysis in this study. Fig. 6 exhibits the misfit stress for large debond energy after neutron irradiation up to 1dpa at 1000°C, obtained by the hysteresis loop analysis methodology. Misfit stress is the reference of residual stress as defined in Ref. [24], and the conversion of misfit stress to residual stress is given elsewhere[23]. Large negative misfit stress corresponds to axial tensile stress in fiber, axial compressive stress in matrix and radial tensile stress at the interphase. Thus this misfit stress is also roughly in agreement with the results of four phase model as shown in Fig. 5 (b).

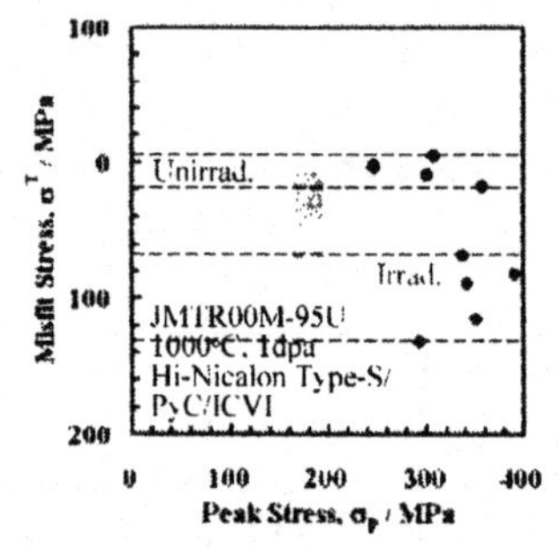

Fig. 6 Neutron irradiation effect on misfit stress after irradiation up to 1dpa at 1000°C, obtained by the hysteresis loop analysis methodology. Misfit stress at highest peak stress of each sample was plotted.

The increase of axial compressive stress in matrix (i.e. negative misfit stress) after irradiation can contribute to the increase of PLS, as described in Eq. (6). Generally, PLS of unidirectional composites, i.e. matrix cracking stress (σ_{mc}), is closely dependent on interfacial properties and modulus of fiber and matrix, and misfit stress [25].

$$\sigma_{mc} = E_c\left[\frac{6\tau\Gamma_m V_f^{\,2} E_f}{(1-V_f)E_m^2 r E_c}\right]^{1/3} - \sigma^T \qquad (6)$$

where σ^T is the misfit stress, Γ_m is the matrix fracture energy, E_f is the Young's modulus of fiber, E_m is that of matrix, E is that of composites, τ is the interfacial sliding stress, V_f is the fiber volume fraction, and r is the fiber radius, respectively. Young's modulus of CVD-SiC[26] and HNLS fiber[27] was degraded, but not significantly after neutron irradiation. Interfacial sliding stress might be nearly the same or degraded slightly after neutron irradiation. Therefore, axial compressive residual stress might be attributed to the retention of PLS after neutron irradiation, even though the composites had a thick PyC interphase.

In addition, the residual stress change by irradiation might cause the increase of UTS. According to Curtin[28], ultimate tensile strength in unidirectional composites depends on the interfacial property,

$$\sigma_u = V_f \sigma_c \left(\frac{2}{m+2}\right)^{\frac{1}{m+1}} \left(\frac{m+1}{m+2}\right)$$
$$\sigma_c = \left(\frac{\sigma_0^m \tau L_0}{r}\right) \tag{7}$$

where σ_u is the ultimate tensile strength, σ_c is the characteristic strength, σ_0 is the Weibull mean strength, m is the Weibull modulus, and L_0 is the gauge length, respectively. The primary role of interphase is to control the fracture behaviors (brittle or quasi-ductile). The maximum strength of composites significantly depends on the performance of intact fibers. Compared with the case of unirradiated SiC/SiC composites, bebonding of PyC interphase/matrix might be enhanced after neutron irradiation, due to the wider difference of axial stress between fiber and matrix, and due to similar axial tensile stress of fiber and PyC interphase, as shown in Fig. 5 (b). Additionally, low fluence of neutron irradiation probably caused slight degradation in interfacial sliding stress, enhancing the deflection of cracks. Remarkable degradation of strength in β-SiC[29, 30] and HNLS fiber[27] was not obtained. From these aspects, PyC interphase/matrix debonding can be enhanced, and deflected cracks can propagate intricately, resulting in the fullest potential of fiber strength. (But, further SEM observations of PyC interphase in the irradiated composites will be needed.)

Therefore, the increase of residual stress induced by the differential swelling between fiber, matrix and PyC interphase may affect PLS and UTS after irradiation up to 1dpa at 1000°C.

SUMMARY

The effects of irradiation-induced swelling of constituents on tensile properties of advanced SiC/SiC composites were examined. After the swelling of the fiber and matrix ion-irradiated up to 1dpa at 1000°C was measured, the four phase model was applied in order to evaluate the residual stress in each constituent. The main results obtained are summarized as follows:

(1) The differential swelling between HNLS and ICVI matrix was obtained (HNLS: 0.27%, ICVI-SiC: 0.48%), while such remarkable mismatch was not detected between TySA (0.46%) and ICVI matrix.

(2) In the four phase model analysis, the residual stresses in the composites were strongly dependent on both CTE and swelling mismatch. Axial tensile stress in fiber, axial compressive stress in matrix, and tensile radial stress at the PyC interphase increased by swelling mismatch of fiber, matrix and PyC interphase after irradiation up to 1dpa at 1000°C.

(3) These increases of the residual stresses were roughly consistent with the tensile stress-strain behavior after neutron irradiation up to 1dpa at 1000°C and the behavior of misfit stress estimated by the hysteresis loop analysis. These results suggest that the increase of residual stress caused by the differential swelling of each constituent might have an affect on PLS and UTS.

ACKNOWLEDGEMENTS

The authors would like to thank Dr. H. Kishimoto and Mrs. O. Hashitomi, S. Kondo, and S. Ikeda for the assistance of ion-irradiation experiments in DuET facility.

This study was financially supported by Development of Innovative Nuclear Technologies by Collaboration among Industrial, Academic and Governmental Circles.

REFERENCES

[1]A.R. Raffray, R. Jones, G. Aiello, M. Billone, L. Giancarli, H. Golfier, A. Hasegawa, Y. Katoh, A. Kohyama, S. Nishio, B. Riccardi and M.S. Tillack, "Design and Material Issues for High Performance SiC_f/SiC-Based Fusion Power Cores," *Fus. Eng. Des.*, **55**, 55-95 (2001).

[2]L.L. Snead, D. Steiner, and S.J. Zinkle, "Measurement of the effect of radiation damage to ceramic composite interfacial strength," *J. Nucl. Mater.*, **191-194**, 566-570 (1992).

[3]L.L. Snead, Y. Katoh, A. Kohyama, J.L. Bailey, N.L. Vauhn, and R.A. Lowden, "Evaluation of neutron irradiated near-stoichiometric silicon carbide fiber composites," *J. Nucl. Mater.*, **283-287**, 551-555 (2000).

[4]M. Takeda, A. Urano, J. Sakamoto, and Y. Imai, "Microstructure and oxidative degradation behavior of silicon carbide fiber Hi-Nicalon type S," *J. Nucl. Mater.*, **258-263**, 1594-1599 (1998).

[5]T. Ishikawa, Y. Kohtoku, K. Kumagawa, T. Yamamura, and T. Nagasawa, "High-strength alkali-resistant sintered SiC fibre stable to 2,200 degrees C," *Nature*, **391**, 773-775 (1998).

[6]T. Hinoki, L.L. Snead, Y. Katoh, A. Hasegawa, T. Nozawa, and A. Kohyama, "The effect of high dose/high temperature irradiation on high purity fibers and their silicon carbide composites," *J. Nucl. Mater.*, **283-287**, 1157-1162 (2002).

[7]T. Nozawa, K. Ozawa, S. Kondo, T. Hinoki, Y. Katoh, L.L. Snead, and A. Kohyama, "Tensile, Flexural and Shear Properties of Neutron Irradiated SiC/SiC Composites with Different F/M Interfaces," *J. ASTM Int.*, **2**, 12884-1-13 (2005).

[8]K. Ozawa, T. Hinoki, T. Nozawa, Y. Katoh, Y. Maki, S. Kondo, S. Ikeda, and A. Kohyama, "Evaluation of Fiber/Matrix Interfacial Strength of Neutron Irradiated SiC/SiC Composites Using Hysteresis Loop Analysis of Tensile Test," *Mater. Trans.*, 207-210 (2006).

[9]C.H. Henager Jr., E.A. Le, and R.H. Jones, "A model stress analysis of swelling in SiC/SiC composites as a function of fiber type and carbon interphase structure," *J. Nucl. Mater.*, **329-333**, 502-506 (2004).

[10]T. Hinoki, L.L. Snead, E. Lara-Curzio, Y. Katoh, and A. Kohyama, "Tensile Testing of Unidirectional Silicon Carbide Composites for Future Irradiation Experiments," *DOE/ER-***0313/29** 74-84 (2000).

[11]A. Kohyama, Y. Katoh, M. Ando, and K. Jimbo, "A new Multiple Beams–Material Interaction Research Facility for radiation damage studies in fusion materials," *Fusion Eng. Des.*, **51-52**, 789-795 (2000).

[12]J.F. Ziegler, J.P. Biersack, and U. Littmark, "The Stopping and Range of ions in Solids," Pergamon Press, 109 (1985).

[13]Y. Katoh, H. Kishimoto, and A. Kohyama, "Low Temperature Swelling in Beta-SiC Associated with Point Defect Accumulation," *Mater. Trans.*, **43**, 612-616 (2002).

[14]K. Ozawa, S. Kondo, H. Kishimoto, and A. Kohyama, "Application of FIB with micro pick-up to microstructural study of irradiated SiC/SiC composites," *J. Electron Microsc.*, **53**, 519-521 (2004).

[15]Y. Mikata, and M. Taya, "Stress Field in a Coated Continuous Fiber Composite Subjected to Thermo-Mechanical Loadings," *J. Compos. Mater.*, **19**, 554-578 (1985).

[16]M. Kuntz, B. Meier, and G. Grathwohl, "Residual Stresses in Fiber-Reinforced Ceramics due to Thermal Expansion Mismatch," *J. Am. Ceram. Soc.*, **76**, 2607-2612 (1993).

[17]C.M. Warwick, and T.W. Clyne, "Development of composite coaxial cylinder stress analysis model and its application to SiC monofilament systems," *J. Mater. Sci.*, **26**, 3817-3827 (1991).

[18]Y. Katoh, H. Kishimoto, and A. Kohyama, "The influences of irradiation temperature and helium production on the dimensional stability of silicon carbide," *J. Nucl. Mater.*, **307-311**, 1221-1226 (2002).

[19]S. Kondo, T. Hinoki, and A. Kohyama, "Synergistic Effects of Heavy Ion and Helium Irradiation on Microstructural and Dimensional Change in β-SiC," *Mater. Trans.*, **46**, 1388-1392 (2005).

[20]J.L. Kaae, "The mechanical behavior of biso-coated fuel particles during irradiation. Part I: Analysis of stresses and strains generated in the coating of a biso fuel particle during irradiation," *Nucl. Technol.*, **35**, 359-367 (1977).

[21]J.L. Kaae, R.E. Bullock, C.B. Scott, and D.P. Harmon, "The mechanical behavior of biso-coated fuel particles during irradiation. Part II: prediction of biso particle behavior during irradiation with a stress-analysis model," *Nucl. Technol.*, **35**, 368-378 (1977).

[22]A.G. Evans, J.-M. Domergue, and E. Vagaggini, "Methodology for Relating the Tensile Constitutive Behavior of Ceramic-Matrix Composites to Constituent Properties," *J. Am. Ceram. Soc.*, **77**, 1425-1430 (1994).

[23]E. Vagaggini, J.-M. Domergue, and A.G. Evans, "Relationships between Hysteresis Measurements and the Constituent Properties of Ceramic Matrix Composites: I, Theory," *J. Am. Ceram. Soc.*, **78**, 2709-2720 (1995).

[24]J.-M. Domergue, E. Vagaggini, and A.G. Evans, "Relationships between Hysteresis Measurements and the Constituent Properties of Ceramic Matrix Composites: II, Experimental Studies on Unidirectional Materials," *J. Am. Ceram. Soc.*, **78**, 2721-2731 (1995).

[25]A.G. Evans, and F.W. Zok, "The physics and mechanics of fibre-reinforced brittle matrix composites," J. Mater. Sci., **29**, 3857-3896 (1994).

[26]M.C. Osborne, J.C. Hay, L.L. Snead, and D. Steiner, "Mechanical- and Physical-Property Changes of Neutron-Irradiated Chemical-Vapor-Deposited Silicon Carbide," *J. Am. Ceram. Soc.*, **82**, 2490-2496 (1992).

[27]T. Nozawa, T. Hinoki, Y. Katoh, and A. Kohyama, "Neutron irradiation effects on high-crystallinity and near-stoichiometry SiC fibers and their composites," *J. Nucl. Mater.*, **329-333**, 544-548 (2004).

[28]W.A Curtin, "Theory of Mechanical Properties of Ceramic-Matrix Composites," *J. Am. Ceram. Soc.*, **74**, 2837-2845 (1991).

[29]R.J. Price, "Effects of fast-neutron irradiation on pyrolytic silicon carbide," *J. Nucl. Mater.*, **33**, 17-22 (1969).

[30]R.J. Price, and G.R. Hopkins, "Flexural strength of proof-tested and neutron-irradiated silicon carbide," *J. Nucl. Mater.*, **108-109**, 732-738 (1982).

BEHAVIORS OF RADIOLUMINESCENCE OF OPTICAL CERAMICS FOR NUCLEAR APPLICATIONS

T. Shikama, S. Nagata, K. Toh, B. Tsuchiya, and A. Inouye
Institute for Materials Research, Tohoku University,
2-1-1 Katahira, Aoba-ku, Sendai 980-8577, Japan
Tel.: +81-22-215-2060, Fax: +81-22-215-2061, E-mail:shikama@imr.tohoku.ac.jp

ABSTRACT

Optical ceramics such as fused silica and transparent alumina are expected to play an important role in nuclear systems under development. For example, optical diagnostics of burning fusion plasma will be crucial for success of the ITER. The paper will describe behaviors of radiation induced luminescence of optical ceramics under gamma-rays, ions, and 14MeV neutron irradiation. Radiation induced luminescence will be sometimes hazardous for optical diagnostics, but it will be also beneficial as a radiation-dosimetry-sensor. Fused silica (SiO_2) emits strong radioluminescence peaks in a visible wavelength range, and behaviors of radioluminescence peaks were found to be strongly modified by existence of oxyhydrate (OH). Alumina (Al_2O_3) is another candidate material for optical windows and it usually has a strong radioluminescence peak at about 693nm, which is caused by the impurity chromium. The radioluminescence peak at 693nm of chromium doped alumina (ruby) is well-established peak for application to the radiation dosimetry, but its intensity changed complicated with changes of irradiation doses, irradiation temperatures and energies of incident ions. Some ceramics, such as strontium aluminates showed radioluminescence under the 14MeV neutron irradiation, and they can be used as a sensor for the 14MeV fusion neutrons.

1. INTRODUCTION

Many ceramics reveal a scintillating behavior in the wavelength range from ultraviolet to infrared region, under irradiation by energetic photons, ions, and neutrons, which is called radiation induced luminescence, being abbreviated as radioluminescence. Radioluminescence is hazardous for optical windows in nuclear fusion plasma diagnostic systems; in the meantime, it could be applied to radiation-dosimetry in nuclear systems. An optical dosimetry system, being composed of radioluminescent ceramics as a radiation-sensor and optical fibers as an optical signal guide, has many attractive features, such as being compact, simple, flexible, robust, and free from need of an outer power supply. It can be deployed in narrow and geometrically complicated places such as coolant channels in a nuclear fission reactor core and in blanket regions in a nuclear fusion reactor.[1]

The optical dosimetry system is a strong candidate, especially in harsh environments where a conventional electrical dosimetry system cannot be applied.[2] An example will be a high temperature and a high electronic-excitation-dose rate radiation environment, where an appropriate electrical insulating material cannot be found for an electrical system. Also, it should be pointed out that the optical system is insensitive to electrical noises which are abundant in nuclear systems. However, vulnerability of optical fibers in radiation environments, namely coloring and losing optical transmissivity, has inhibited the optical dosimetry system from being deployed in nuclear systems until recently. Recent improvement of radiation-resistance of fused silica (SiO_2) core optical fibers is making the optical dosimetry system a realistic choice.[3-5]

Several features are demanded for ceramic scintillators as a radiation monitor, such as a good radiation resistance, a wide dynamic range, and refractoriness. Also, a different response to a different irradiation will be important to discriminate kinds of radiation sources. In principle, the direct cause of radioluminescence is the electronic excitation effects of the radiation. An electron is excited by the radiation to a specific electronic level and then drops into a lower electronic level, emitting photons. However, an electron level which is responsible for the radioluminescence is usually associated with a lattice defect, which will be created by the atomic displacement effects of the radiation, and with an impurity whose chemical and electrical states will be affected by the radiation effects. Also, behaviors of electron holes have an important role, which will be affected by the radiation effects.

In the present paper, behaviors of some luminescent materials will be described under gamma-rays, ion-beams, and neutron irradiation. Radioluminescent behaviors of these ceramics were found to depend on their detailed response to specific radiation, namely, formation of lattice point defects due to the atomic displacement effects, and formation of electronic defects due to the electronic excitation effects and the radiolysis effects. Appropriate choice of radioluminescent material is found to enable dosimetry of a radiation dose rate, an accumulated radiation dose and energy of radiation as well as discrimination of radiation species.

2. EXPERIMENTS

Radioluminescent behaviors of the ceramics, the silica (SiO_2), the ruby (chromium doped alpha alumina (α-Al_2O_3)) and strontium aluminates doped with europium and dysprosium ($SrAl_2O_4:Eu^{2+},Dy^{3+}$ and $Sr_4Al_{14}O_{25}:Eu^{2+},Dy^{3+}$) were studied under the Co^{60} gamma-ray irradiation, hydrogen and helium ions of 0.1-2.0MeV, and the 14MeV fusion neutrons. The Co^{60} gamma-ray irradiation was carried out in an ambient air in the temperature range of 300-600K, at the Takasaki Research Establishment of the Japan Atomic Energy Agency (JAEA). The maximum dose rate is about 7Gy/s. The hydrogen and helium ion irradiations were done in a tandem-type accelerator of Institute for Materials Research of Tohoku University in a vacuum of about 2×10^{-6}Pa in the temperature range of 300-800K. The ion flux is about 6×10^{12}ions/m^{2}s. The 14MeV fast neutron irradiation was performed in an ambient air at the deuterium-tritium neutron irradiation facility of the fusion neutronics source (FNS) at the Tokai Research Establishment of the JAEA. The energy spectrum of the neutrons exhibits a sharp peak at 14 MeV and the flux was in the range of 10^6 to 10^9 n/cm^{2}s. The associated gamma-ray dose rate was marginal, being about 10^{-2}Gy/s.

Geometrical configurations of an irradiation setup are different among different irradiations and a schematic diagram of the irradiation setup for the ion irradiation is shown in Fig. 1. Detailed setups for other irradiations can be found elsewhere.[3, 6-10] In general, radioluminescence from irradiated materials were guided through a radiation resistant fused silica optical fiber to a photonic multichannel analyzer (Hamamatsu Photonics, PMA-10 and PMA-11). The measuring wavelength range is 300-800nm.

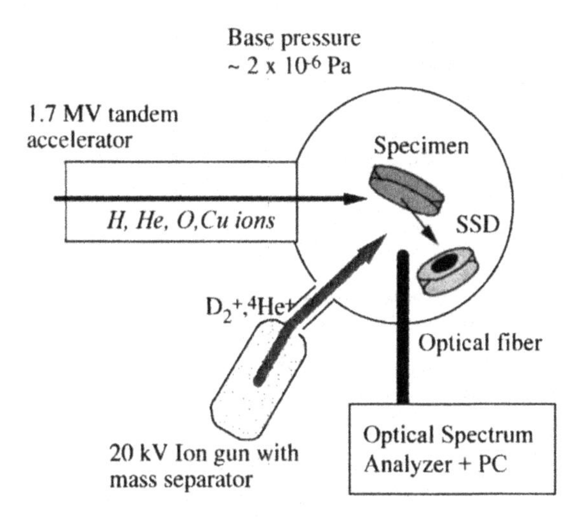

Figure 1 Experimental setup for ion irradiation[10]

3. RESULTS AND DISCUSSIONS

(a) fused silica [9-11]

Radioluminescence is, in general, emitted by a specific electron level associated with a specific lattice defect. Figure 2 (a) shows radioluminescence spectra of fused silica containing different oxyhydrate (OH) concentrations, 0, 200 and 800ppm, at the onset of 1.5MeV proton irradiation. Radioluminescence peaks at 3.1eV (400nm), 2.7eV (460nm) and 1.9eV (650nm) are assigned as B2β, B2α, and NBOHC (Non Bridging Oxygen Hole Center), respectively. The B2β radioluminescence is emitted by an intrinsic lattice defect (oxygen deficit structure being introduced in the course of manufacturing processes) which is unstable under the ionizing irradiation and is easily converted into a different defect of B2α. Also, the OH doping cures the lattice defects responsible for the B2α and the B2β, which results in a very low radioluminescence intensity in the fused silica containing 800ppm OH at the onset of irradiation. However, radiation effects introduce the lattice defect responsible for the B2α and a rapid growth of the B2α radioluminescence is observed in all the fused silica as shown in Fig. 2(b).

Intensity of the B2α increases with the ion fluence, roughly depending on the square root of the ion fluence. Thus, the intensity of the B2α could be a measure to estimate an ion fluence. In the meantime, the growth of the NBOHC saturated at an early stage of the irradiation, at about 1×10^{18} protons/m^2, and the intensity of the NBOHC could be a measure for estimating an ion flux. Then, the combination of the B2α and the NBOHC could yield information about a flux and a fluence of the concerned ion radiation.

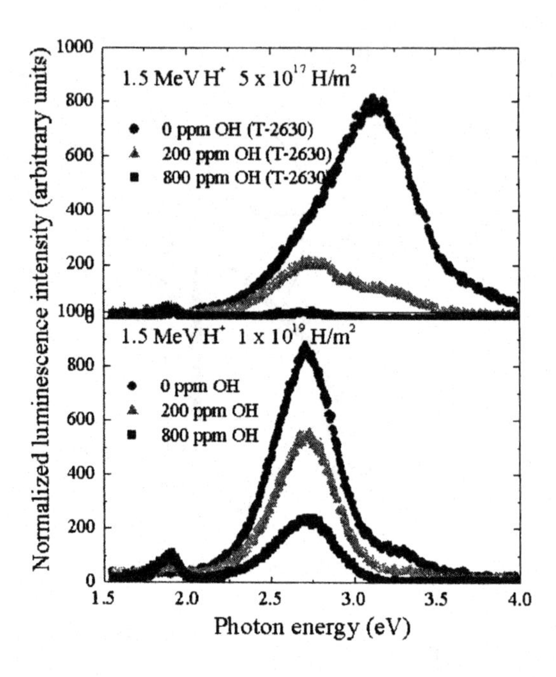

Figure 2 Radioluminescence of fused silica. (a) at the start of 1.5MeV proton irradiation; 5×10^{17}proton/m^2, (b) at prolonged irradiation, 1×10^{19}proton/m^2. [11]

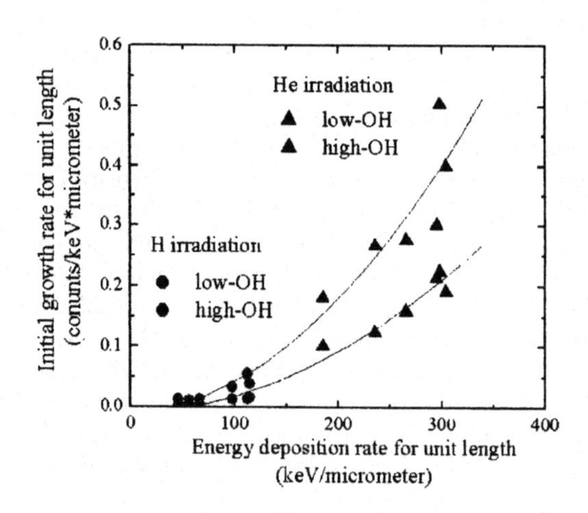

Figure 3 Initial growth rate of B2α radioluminescence peak at 460nm in fused silica as a function of electronic energy deposition rate for unit length.[11]

The growth behavior of the B2α is not governed by the atomic displacement effects but by the electronic excitation effects of ion radiation. The growth rate of the B2α does not show good correlation with the atomic displacement rate but show the clear dependence on the electronic excitation rate as shown in Fig. 3. This means that the lattice defects responsible for the B2α are generated by the electronic excitation processes, namely, a radiolysis is taking place in the fused silica. Thus, the behaviors of radioluminescence of the fused silica are sensitive to the electronic excitation effects of the concerned radiation.

(b) Ruby[12]

Chromium doped alpha-alumina (the ruby) is a well-established radioluminescent material, which emits 693nm luminescence from the ^2E electron level associated with the chromium 3+ ion (Cr^{3+}) in the alumina. Intensity of the radioluminescence of ruby at 693nm has a good linear dependence on the electronic excitation dose rate, in the range of electronic excitation dose rate smaller than a few 100Gy/s. Above a few 100Gy/s dose rate, the intensity of 693nm radioluminescence showed a saturation behavior with further increase of the electronic excitation dose rate. This saturation behavior will be due to a long life time of the excited electron state corresponding to 693nm emission, namely about 3 milliseconds, which will be a dead time for each electron level.

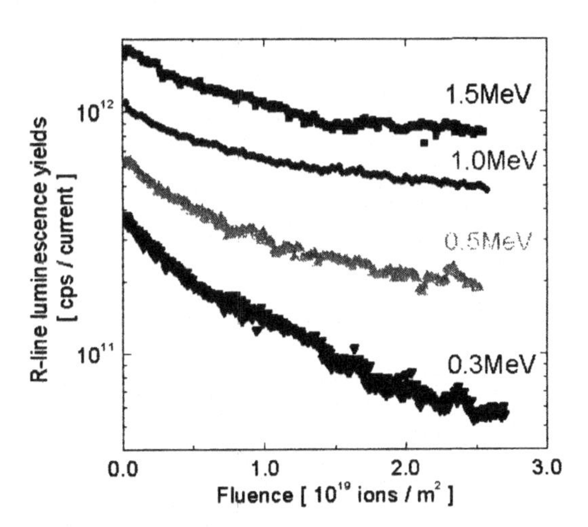

Figure 4 Intensity of 603nm radioluminescence peak of ruby as a function of helium ion fluence.[12]

Under a prolonged ion radiation, intensity of the 693nm radioluminescence of the ruby decreased with the increase of the ion fluence as shown in Fig. 4. Several causes were proposed for this decrease of intensity of the 693nm radioluminescence; change of a chromium electrical charge state, agglomeration of chromium ions, diffusion of chromium ions to grain boundaries, etc.. Figure 5 shows that a decreasing rate of the

693nm radioluminescence had a good correlation with a atomic displacement rate but not with an electronic excitation dose rate. Also, growth of the F-center radioluminescence at 420nm was not observed in the ruby, suggesting that the oxygen vacancies had some interaction with some defects. These two experimental results imply that the Cr^{3+}, which has an ion diameter larger than that of Al^{3+} ion and will have a compressive stress field around it, will have a strong positive interaction with the oxygen vacancy and forms a Cr^{3+}-(oxygen vacancy) complex. Recovery of the once-decreased 693nm radioluminescence in the course of subsequent annealing took place at about 400-600C in case of the 1MeV proton irradiation of $5x10^{20}$proton/m^2, which will suggest that the oxygen vacancy trapped by the Cr^{3+} is released and become mobile in this temperature range. With larger ion fluences, another recovery stage was observed above 800C, especially in the case of helium ion irradiation. In this case, dissociation of once agglomerated of Cr^{3+}-(oxygen vacancy) complexes will be responsible for the recovery.

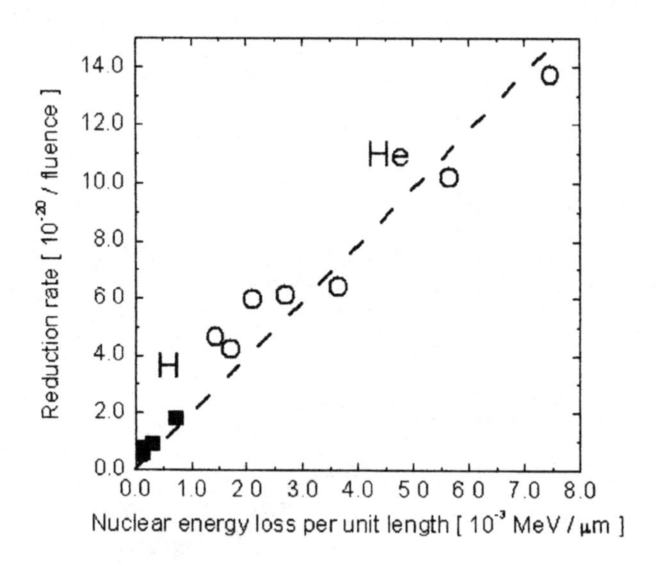

Figure 5 Initial reduction rate of ruby 693nm radioluminescence peak as a function of nuclear energy loss per unit length.[12]

In principle, intensity of the 693nm radioluminescence of the ruby is a good monitor for measuring the electronic excitation dose rate, in the dose rate smaller than a few 100Gy/s. However, above a few 100Gy/s, its sensitivity decreased with increase of the electronic excitation dose rate. Also, the atomic displacement effects decreased its sensitivity. In the meantime, the decrease rate of its sensitivity will be a good measure for evaluating a displacement dose rate of the concerned radiation.

(c) Strontium aluminate doped with europium and dysprosium [6,8,13]

As described above, radioluminescence is generated in principle by the electronic excitation process. Then, detection of neutrons by the radioluminescence is difficult as the neutron has very weak direct electronic excitation effects. Under the 14MeV neutron irradiation with the present neutron fluxes, radioluminescence from the fused silica and the ruby could not be detected. Thus, search for ceramics which will emit radioluminescence under neutron radiation is important. Strontium aluminates doped with europium and dysprosium ($SrAl_2O_4:Eu^{2+},Dy^{3+}$ and $Sr_4Al_{14}O_{25}:Eu^{2+},Dy^{3+}$, hereafter denoted as strontium aluminate) are found to emit strong radioluminescence under the 14MeV fusion neutron radiation. Figure 6 shows radioluminescence spectra from the strontium aluminates, under the 14MeV fusion neutron radiation. Strong radioluminescence peaks (complex of peaks) are observed at about 500nm, whose intensities are nearly proportional to the neutron flux. However, the present strontium aluminates are so-called long lasting phosphors (LLPs) and they continue to emit luminescence after the irradiation stops, with the decrease of their intensity with time. Thus, they are not a suitable detector for monitoring a neutron flux which is rapidly changing.

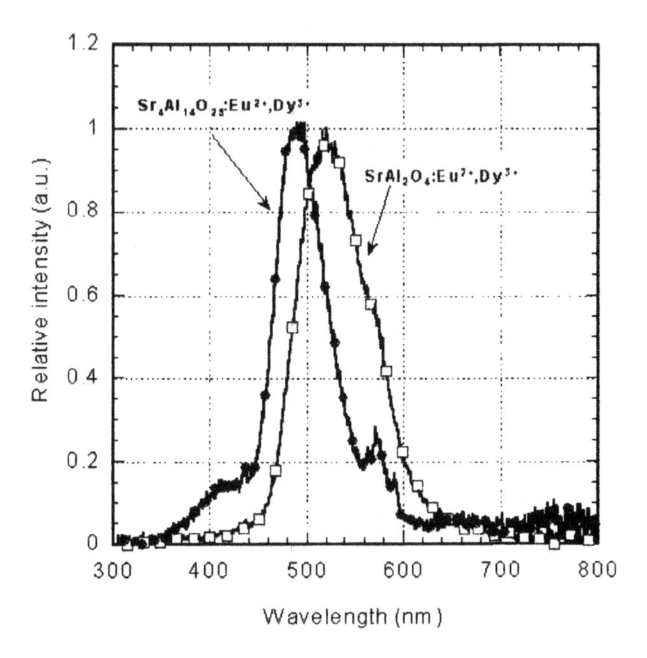

Figure 6 Main radioluminescence peak at about 500nm from strontium aluminateds doped with europium and dysprosium, under 14MeV neutron irradiation.[13]

In the case of the $Sr_4Al_{14}O_{25}:Eu^{2+},Dy^{3+}$, major luminescence peaks (complex of peaks) are at about 400nm, about 480nm, and about 570nm. The luminescent peaks at 400 and 480 nm are attributed to Eu^{2+}, while the peak at 570 nm is attributed to Dy^{3+}. The peaks (complexes of peaks) at 400nm and 480nm are long lasting phosphorescence, due to the interaction of Eu^{2+} with electron holes, but the complex of peaks at about 570nm disappeared promptly when the 14MeV neutron radiation stopped. So, a fast change of the 14MeV neutron flux could be monitored by the intensity of 565nm radioluminescence peak (complex of peaks).

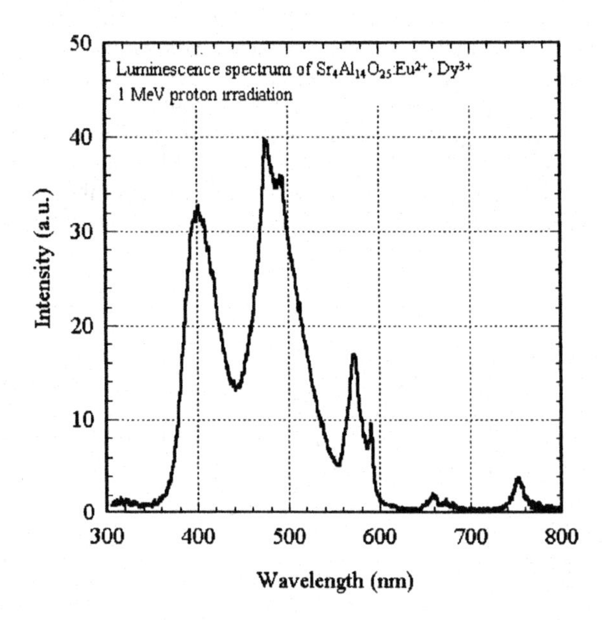

Figure 7 Radioluminescence spectra of $Sr_4Al_{14}O_{25}$;Eu^{2+},Dy^{3+} under 1MeV proton irradiation.[13]

Spectra of the radioluminescence of $Sr_4Al_{14}O_{25}$:Eu^{2+},Dy^{3+} have an interesting dependence on species of radiation and on the ion energy. Figure 7 shows the radioluminescence spectrum observed under the 1MeV protons, which is different from that observed under the 14MeV neutrons. So, discrimination of the 14MeV neutron flux from ion influxes is possible by measuring a detailed spectrum of the radioluminescence. Also, energy of incident ions could be evaluated by measuring a detailed spectrum of the radioluminescence. Figure 8 shows ratios of the intensities of radioluminescence peaks (complexes of peaks), namely 480nm/570nm and 400nm/570nm. The ratios have clear dependence on the proton energy.

Intensity of the major radioluminescent peaks of $Sr_4Al_{14}O_{25}$:Eu^{2+},Dy^{3+} at 480nm and 570nm did not show clear decrease in the course of the 14MeV neutrons up to the neutron fluence of $10^{18}n/m^2$. The peak at 480nm is attributed to Eu^{2+}, which is occupying a Sr^{2+} site in the lattice and will be relatively stable due to a smaller size- misfit. In the meantime, the radioluminescence peak at 570nm, which responded promptly to a prompt change of the neutron flux, is attributed to Dy^{3+}, which is occupying a Al^{3+} site and has a larger size-misfit. So, the decrease of the intensity in the prolonged irradiation should be expected, similar to that observed in the ruby, due to the interaction with oxygen vacancy and the agglomeration.

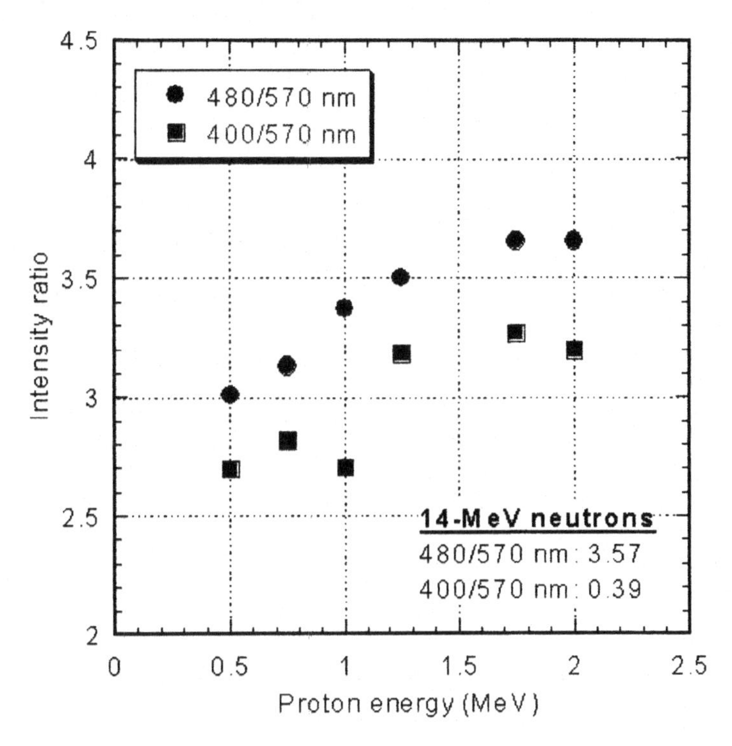

Figure 8 Intensity ratio of radioluminescence peaks in $Sr_4Al_{14}O_{25}$:Eu^{2+},Dy^{3+}, under helium ion irradiation.[13]

4. CONCLUSIONS

Behaviors of radioluminescence of the fused silica (SiO2), the chromium doped a-alumina (Al_2O_3:Cr^{3+}), and the strontium aluminates doped with europium and dysprosium ($SrAl_2O_4$:Eu^{2+},Dy^{3+} and $Sr_4Al_{14}O_{25}$:Eu^{2+},Dy^{3+}) were studied under the irradiation with Co^{60} gamma-ray, proton and helium ions, and the 14MeV fusion neutrons. Measuring not only the intensity of the radioluminescence but also the changing rate of the intensity will yield the information about the radiation flux as well as the radiation fluence. Also, the electronic excitation effects and the atomic displacement effects could be distinguished through detailed analysis of the behaviors of radioluminescence spectra. The 14MeV neutron dosimetry, which is important but difficult, is possible with the strontium aluminates.

REFERENCES

[1]S. Yamamoto, T. Shikama, V. Belyakov, E. Farnum, E. Hodgson, T. Nishitani, D. Orlinski, S. Zinkle, S. Kasai, P. Stott, K. Young, V. Zaveriaev, A. Costley, L. deKock, C. Walker, G. Janeschitz, J. Nucl. Mater., 283–287 (2000) 60–69.
[2]T. Shikama, K. Yasuda, S. Yamamoto, C. Kinoshita, S.J. Zinkle, E.R. Hodgson, J. Nucl. Mater., 271&272 (1999) 560–568.
[3]T. Shikama, T. Kakuta, N. Shamoto, M. Narui, T. Sagawa, Fusion Eng. Des., 51–52 (2000) 179–183.
[4]A.A. Ivanov, S.N. Tugarinov, Yu.A. Kaschuck, A.V. Krasilnikov, S.E. Bender, Fusion Eng. Des., 51–52 (2000) 973–976.
[5]A.L. Tomashuk, V.A. Bogatyrjov, E.M. Dianov, K.M. Golant, S.N. Klyamkin, I.V. Nikolin, M.O. Zabezhailov, Proceedings of the SPIE, 4547 (2002) 69–73.
[6]T. Shikama, K. Toh, S. Nagata, B. Tsuchiya, M. Yamauchi, T. Nishitani, T. Suzuki, K. Okamoto, N. Kubo, Nucl. Fusion 46 (2006) 46-50.
[7]K.Toh, T.Shikama, S.Nagata, B.Tsuchiya, T.Kakuta, T.Hoshiya, M.Ishihara, Fusion. Sci.Technol, 44(2) (2003) 475-476.
[8]T. Shikama, K. Toh, S. Nagata, and B.Tsuchiya, Measurement Science and Technology (2006) in press.
[9]S.Nagata, S. Yamamoto, K. Toh, B. Tsuchiya, T. Shikama, H. Naramoto, "Ion induced luminescence of silica glasses and optical fibers", in proceedings of SPIE2003, SPIE-5199 (2004), San Diego, August, 2003.
[10]S.Nagata, S. Yamamoto, K. Toh, B. Tsuchiya, T. Shikama, H. Naramoto, J. Nucl. Mater., 329-333 (2004) 1507-1510.
[11]S.Nagata, presented at the 12th Int. Conf. on Fusion Nuclear Materials, Santa Barbara, December, 2005.
[12]A.Inouye, S.Nagata, et al., presented at the 12th Int. Conf. on Fusion Nuclear Materials, Santa Barbara, December, 2005.
[13]K.Toh et al., presented at the 12th Int. Conf. on Fusion Nuclear Materials, Santa Barbara, December, 2005.

Author Index